Descriptive Handbook
of the Rock Forming Minerals

Rob Kanen BSc

ISBN 1450576842

Table of Contents

Introduction and Terminology

The Descriptive Handbook of Rock Forming Minerals is a reference book of all rock forming and accessory minerals. There are two listings for each mineral: 1. Physical properties, listing all physical properties and 2. Optical properties, listing all optical properties visible with a polarizing microscope. Where minerals are of a small size or microscopic in nature, only the optical properties are listed for that mineral.

The following terms are used in the mineral descriptions:

Physical Properties

Formula – The chemical formula

Crystal System - The crystal system
Isometric or cubic
Hexagonal or trigonal
Tetragonal
Orthorhombic
Monoclinic
Triclinic

Group - A mineral group name
Quartz
Pyroxene
Feldspar
Garnet
Serpentine
Clay
Mica
Brittle Mica

Bauxite

Oxide

Amphibole

Feldspathoid

Zeolite

Silliminite

Calcite

Barite

Humite

Tourmaline

Epidote

Chlorite

Specific Gravity - The relative weight of a mineral compared to water

Hardness - Hardness of a mineral according to Moh's Hardness Scale

1 - Soft, i.e. Talc

2 - Copper

3 - Calcite

4 - Fluorite

5 - Medium i.e. Apatite

6 - Feldspar

7 - Glass or Quartz

8 - Topaz

9 - Corundum

10 - Hardest natural occurring mineral, Diamond

Cleavage - Cleavage outlines are often distinctive of a mineral.

Cubic Cleavage - Galena

Cleavage/Foliation Masses/Aggregates - Aggregates of planar minerals, such as chlorite and mica in one direction, parallel to cleavage.

Streak - Color of a minerals powder when crushed. Determined by rubbing a mineral on a porcelain plate.

White - Most common, feldspar, calcite, garnet

Black - Magnetite, uraninite

Green - Hornblende, vesuvianite

Color - Natural color of a mineral

Black - Mica, Uraninite, Hematite

White - Plagioclase

Pink - Orthoclase

Green - Chlorite, olivine

Opacity - The transparency of a mineral

Opaque - Does not transmit light

Translucent - Partially transmits light

Transparent - Fully transmits light

Luster - Appearance of a minerals surface

Pearly - Smooth, shiny, white surface, talc, calcite

Vitreous - Glass like, most transparent minerals are vitreous, quartz, garnet

Dull - Does not shine, talc, kaolinite

Adamantine - Brilliant shining surface, diamond

Soapy - Soapy feel, talc, chlorite

Greasy - Greasy feel, topaz, olivine

Silky - Like silk, antigorite, anthophyllite

Waxy - Wax like surface, quartz, serpentine,

Satin - Satin like surface, kaolinite

Submetallic - Shiny, opaque, metallic like surface, mica

Metallic - Shiny, metallic surface, pyrite

Splintery - In splinters, chlorite

Habit - A minerals external form

Cubic - In cubes, pyrite, halite garnet

Polyhedrons/Octohedrons - Garnet, fluorite

Rhombohedral - Rhomb shaped, calcite, chiastolite

Prismatic - Rectangular, square outlines, very common, hornblende, epidote, augite, zircon, sphene

Hexagonal - Hexagonal or triangular outlines, tourmaline, beryl, topaz

Tabular - Thick, flat prismatic outlines, feldspar, biotite, chlorite, epidote, olivine

Columnar - Thin columns, often in aggregates, tourmaline, actinolite/tremolite, hornblende, diopside

Flakes/Plates/Scales - Mostly the fine, microcrystalline varieties. Mica, clay minerals, chlorite, sericite

Fibrous - In fibers, actinolite/tremolite, talc, serpentine, nephrite

Acicular - Fine, needle like crystals, tourmaline, rutile

Radiating - Outwardly radiating from a center, zeolite, tourmaline, gypsum

Spherulitic - A radiating acicular mass of crystals forming a circular shaped pattern. Common pattern in devitrified volcanic glass

Short - Short

Long - Long

Occurrence - The mode of occurrence and/or formation of a mineral

Veins - Hydrothermal veins, quartz, calcite veins

Pegmatites - V.coarse grained veins of granites, feldspars, tourmaline, micas

Vesicular/Amygdaloidal - Cavities in volcanic rocks

Evaporites - Within sedimentary evaporite basins

Metamorphic (Contact) - Along contact zones of recrystallized rocks, hornfels, skarn

Metamorphic (Regional) - Metamorphic rocks occurring over a large area, gneiss

Igneous - Generally some form of magmatic intrusion, granite, gabbro

Sedimentary - Derived from pre-existing rocks and often laid down in layers, sandstone, shale

Sedimentary Limestones - Deposited in ocean basins due reef building or accumulation of carbonate grains (detrital or precipitation), limestone

Optical Properties

Most of these properties are only discernable with a polarizing microscope, as used by geologists.

Relief - The visibility of a mineral in plane polarized light. Usually compared to adjacent minerals or the cementing material used in the slide, such as balsam.

Birefringence - The interference color observable under crossed polars. Refer to a birefringence table to determine values

2V - The angle of the optical axis in biaxial minerals. Measured in an oriented crystal with a Bertrand Lens under crossed polars and high power.

Optical Sign - Most minerals are either uniaxial, with a single optic axis or biaxial, with two optic axes i.e. quartz, calcite, tourmaline are uniaxial, olivine, augite, hornblende, feldspar are biaxial.

Refractive Index - The degree to which a crystal bends light, as it passes through a crystal. The RI varies according to the optical axes. Uniaxial minerals have two directions of RI: 1. Along the ordinary ray; and 2. Along the extraordinary ray.. Biaxial minerals have three directions of RI: 1. Nalpha, 2. Nbeta and 3. Ngamma.

Pleochroism - The variation in color as a mineral is rotated under plane polarized light. I.e. hornblende, biotite are strongly pleochroic

Extinction Angle - The type of extinction when a mineral is rotated under crossed polarized light.
Parallel - Parallel to cleavage or crystal outlines
Oblique - At an angle to cleavage or crystal outlines
Symmetrical - Symmetrical to crystal shape or cleavage, i.e. hornblende, hypersthene, dolomite

Cleavage - Cleavage outlines in thin section are often distinctive of a mineral. There may be some difference between observable cleavages in optical properties and physical properties due to recognition of imperfect cleavages or partings.

Amphibole Cleavage - Two at 56" and 124" in cross section

Pyroxene Cleavage - Two at 83" and 97" in cross section

Mica Cleavage - Perfect in one direction

Chlorite Cleavage - Perfect in one direction

Cubic Cleavage - Halite, galena

Cleavage/Foliation Masses/Aggregates - Aggregates of planar minerals, such as chlorite and mica in one direction, parallel to cleavage.

Twinning - Twinned crystals are often observed in thin section and are distinctive of certain minerals

Polysynthetic Twinning Albite Law - Plagioclase feldspars almost always exhibit this. Used to determine Albite/Anorthite content

Simple Twinning, Amphibole - Hornblende often is twinned with two crystals sharing a common twin plane

Simple Twinning, Carlsbad - Very common in K-feldspars

Penetration Twinning - Where two crystals penetrate each other through the center, andalusite, staurolite, cordierite

Mineral Descriptions

Physical Properties

Name: Acmite (Aegerine)

Group: Pyroxene

Formula: NaFeSi2O6

Crystal System: Monoclinic

Color: Dark green to greenish black, reddish brown

Opacity: Translucent to opaque

Luster: Vitreous to somewhat resinous

Streak: Pale Yellowish grey

SGLow: 3.50

SGHigh: 3.60

HardnessLow: 6

HardnessHigh: 6

Cleavage: 2

Direction: {110} good; {100} parting

Habit: Crystals long prismatic, vertically striated or furrowed; tufts or acicular to capillary.

Fracture: Uneven; brittle

Other: Characteristic of soda-rich igneous rocks, such as phonolite, nepheline- trachyte, soda-granite, soda-syenite

Comments:

Optical Properties

Name: Acmite (Aegerine)

Group: Pyroxene

Formula: NaFeSi2O6

Crystal System: Monoclinic

Color: Green

Form: Long ,bladed prismatic crystals with four or eight sided cross sections.

Relief: High, n>balsam

Birefringence: Strong to v.strong, 0.037-0.059, color of the mineral may mask birefringence

2V: 60"-66"

Nalpha or Nord.: 1.745-1.777

NBeta or Nextr.: 1.770-1.823

NGamma: 1.782-1.836

Optical Sign: Biaxial negative

Orientation: Length fast, r>v

Pleochroism: Strong pleochroism from dark green-light green-yellow

Twinning:

Cleavage: In two directions 87" and 93" and parallel to longitudinal sections

Extinction: Oblique 2"-10" in longitudinal sections

Alteration:

Features: Other pyroxenes have larger extinction angles. Acmite is brown in color.

Occurrence: Characteristic of soda-rich igneous rocks, such as phonolite, nepheline- trachyte, soda-granite, soda-syenite

Physical Properties

Name: Actinolite
Group: Amphibole
Formula: Ca2(Mg,Fe)5Si8O22(OH)2
Crystal System: Monoclinic
Color: Light green to blackish green
Opacity: Transparent to nearly opaque
Luster: Vitreous or dull
Streak: Colorless
SGLow: 3
SGHigh: 3.44
HardnessLow: 5
HardnessHigh: 6
Cleavage: 2
Direction: {110} good, {100} parting
Habit: Crystals long bladed; usually in fibrous to thin columnar aggregates, often radiated; massive
Fracture: Subconchoidal to uneven; brittle
Other: Widespread within metamorphic rocks, especially schists and meta-igneous rocks. Alteration product of pyroxenes and olivine. Also in metamorphosed limestones and skarns
Comments: Series with tremolite & ferro-actinolite

Optical Properties

Name: Actinolite
Group: Amphibole
Formula: Ca2(Mg,Fe)5Si8O22(OH)2
Crystal System: Monoclinic
Color: Colorless to green
Form: Crystals long bladed; usually in fibrous to thin columnar aggregates, often radiated; massive
Relief: High, n> balsam
Birefringence: Moderate to strong, 0.022-0.027
2V: 79"-85"
Nalpha or Nord.: 1.600-1.628
NBeta or Nextr.: 1.613-1.644
NGamma: 1.625-1.655
Optical Sign: Biaxial negative
Orientation: Length fast, r<v weak
Pleochroism: Weak
Twinning: Twinning along {100} common, polysynthetic twinnning along {001} less common
Cleavage: Two {110} at 56" and 124", parallel to length in longitudunal sections
Extinction: Oblique, 10"-20" in longituidinal sections
Alteration: Alters to talc
Features: Extinction, birefringence and amphibole cross section distinctive
Occurrence: Widespread within metamorphic rocks, especially schists and meta-igneous rocks. Alteration product of pyroxenes and olivine. Also in metamorphosed limestones and skarns

Physical Properties

Name: Aegerine Augite
Group: Pyroxene
Formula: $NaFeSi_2O_6$
Crystal System: Monoclinic
Color: Dark green to greenish black, reddish brown
Opacity: Translucent to opaque
Luster: Vitreous to somewhat resinous
Streak: Pale Yellowish grey
SGLow: 3.50
SGHigh: 3.60
HardnessLow: 6
HardnessHigh: 6
Cleavage: 2
Direction: {110} good; {100} parting
Habit: Crystals long prismatic, vertically striated or furrowed; tufts or acicular to capillary.
Fracture: Uneven; brittle
Other: Characteristic of soda-rich igneous rocks, such as phonolite, nepheline- trachyte, soda-granite, soda-syenite

Optical Properties

Name: Aegerine Augite
Group: Pyroxene
Formula: $NaFeSi_2O_6$
Crystal System: Monoclinic
Color: Green
Form: Usually euhedral prismatic crystals
Relief: High, n>balsam
Birefringence: Strong, 0.029-0.037
2V: 60"
Nalpha or Nord.: 1.680-1.745
NBeta or Nextr.: 1.687-1.770
NGamma: 1.709-1.782
Optical Sign: Biaxial negative or positive
Orientation: Length fast, r>v
Pleochroism: Pleochroic from greenish-yellow green
Twinning: Simple twinning along {100} common
Cleavage: In two directions at 87" and 93". Parallel in longitudinal sections.
Extinction: Oblique -15"--38" in longitudinal sections, parallel in cross sections.
Alteration:
Features: Similar to aegerine, but larger extinction angle.
Occurrence: Characteristic of soda-rich igneous rocks, such as phonolite, nepheline- trachyte, soda-granite and soda-syenite

Physical Properties

Name: Albite

Group: Feldspar

Formula: NaAlSi3O8

Crystal System: Triclinic

Color: White to colorless, bluish, grey, reddish, greenish

Opacity: Transparent to subtranslucent

Luster: Vitreous, pearly

Streak: White

SGLow: 2.60

SGHigh: 2.63

HardnessLow: 6

HardnessHigh: 6.5

Cleavage: 2

Direction: {001} perfect, {010} nearly perfect

Habit: Crystals tabular, small; usually massive, lamellar, or granular; twinning common, simple, multiple, and repeated

Fracture: Conchoidal to uneven; brittle

Other: Most abundant in felsic igneous rocks. Widespread occurrence

Comments:

Optical Properties

Name: Albite

Group: Feldspar

Formula: NaAlSi3O8 An0-10%

Crystal System: Triclinic

Color: Colorless

Form: Plates or lathe shaped sections, rarely in phenocrysts

Relief: Low, n considerably <balsam, n=balsam

Birefringence: Weak, 0.009-0.011

2V: 77"-82"

Nalpha or Nord.: 1.527-1.533

NBeta or Nextr.: 1.531-1.537

NGamma: 1.538-1.542

Optical Sign: Biaxial positive

Orientation:

Pleochroism:

Twinning: Polysynthetic twinning according to albite law. Twinning according to carlsbad law either alone or combined. Occasional pericline twinning

Cleavage: Four directions, perfect {001}, distinct {010}, imperfect {110} and {11_0}

Extinction: Maximum in albite twins is 12'-19". In cleavage sections parallel to {001} 3"-5" and parallel to {010} 15"-20"

Alteration: Alters to clay

Features: RI equal to and less than balsam is distinctive, albite twinning and extinction angles help.

Occurrence: Most abundant in felsic igneous rocks. Widespread occurrence

Physical Properties

Name: Allanite-[Ce]
Group: Epidote
Formula: (Ce,Ca,Y)2(Al,Fe)3(SiO4)3(OH)
Crystal System: Monoclinic
Color: Brown to black
Opacity: Translucent to opaque
Luster: Resinous or pitchy submetallic
Streak:
SGLow: 4.12
SGHigh: 4.12
HardnessLow: 5.5
HardnessHigh: 6
Cleavage: 0
Direction:
Habit: Crystals usually tabular; long prismatic to acicular; commonly massive, bladed, or as embedded grains; twinning on {100} common, polysynthetic
Fracture: Conchoidal to uneven; brittle
Other: Occurs within granites, syenites, granite pegmatites and gneisses
Comments:

Optical Properties

Name: Allanite-[Ce]
Group: Epidote
Formula: (Ce,Ca,Y)2(Al,Fe)3(SiO4)3(OH)
Crystal System: Monoclinic
Color: Brown
Form: Elongated prismatic, tabular and bladed crystals, columnar aggregates and six sided cross sections as in epidote
Relief: High, n>balsam
Birefringence: Strong, 0.01-0.03
2V: Large
Nalpha or Nord.: 1.64-1.77
NBeta or Nextr.: 1.65-1.77
NGamma: 1.66-1.80
Optical Sign: Biaxial negative
Orientation: Difficult
Pleochroism: Pleochroic from pale brown - dark brown
Twinning: Infrequent polysynthetic
Cleavage: One, imperfect parallel to {001}
Extinction: Usually parallel
Alteration: Altered to amorphous material with similar composition
Features: Color and pleochroism distinctive. Similar to hornblende, but lacks amphibole cross section, with parallel extinction and one cleavage, instead of two. Cesium variety of epidote.
Occurrence: Occurs within granites, syenites, granite pegmatites and gneisses

Physical Properties

Name: Almandine

Group: Garnet

Formula: Fe3Al2(SiO4)3

Crystal System: Cubic

Color: Deep red, brownish red, brownish black, purplish red

Opacity: Transparent to translucent

Luster: Vitreous to resinous

Streak: White

SGLow: 4.1

SGHigh: 4.3

HardnessLow: 7

HardnessHigh: 7.5

Cleavage: 1

Direction: {110} parting sometimes distinct

Habit: Crystals usually dodecahedrons or trapezohedrons; also in combination or with hexoctahedron; massive; granular

Fracture: Conchoidal to uneven; brittle

Other: Predominantly found in metamorphic rocks, especially schists and gneisses

Comments: Series with pyrope and with spessartine

Optical Properties

Name: Almandine

Group: Garnet

Formula: Fe3Al2(SiO4)3

Crystal System: Isometric, cubic

Color: Colorless to pale reddish

Form: Euhedral dodecahedrons in six sided trapezohedrons in eigth sided cross sections. Plygonal grains and aggregates.

Relief: V. high, n>balsam

Birefringence: Isotropic but mat show weak birefringence

2V:

Nalpha or Nord.: 1.778-1.815

NBeta or Nextr.:

NGamma:

Optical Sign:

Orientation:

Pleochroism:

Twinning:

Cleavage: Parting parallel to {110}, irregular fractures

Extinction:

Alteration:

Features: Similar to spinel, which is octohedral. Determination of RI will differentiate garnets.

Occurrence: Predominantly found in metamorphic rocks, especially schists and gneisses

Optical Properties

Name: Alunite

Group: Alunite

Formula: $KAl_3(OH)_6(SO_4)_2$

Crystal System: Hexagonal

Color: Colorless

Form: Fine to coarse aggregates of tabular to pseudo cubic rhombohedral

Relief: Fair, n>balsam

Birefringence: Moderate, 0.020

2V:

Nalpha or Nord.: 1.572

NBeta or Nextr.: 1.592

NGamma:

Optical Sign: Uniaxial positive

Orientation: Length fast

Pleochroism:

Twinning:

Cleavage: Fair in one direction, {0001}

Extinction: Parallel or symmetrical

Alteration:

Features: Tabular crystals, birefringence

Occurrence: Hydrothermal alteration product of rhyolites, dacites and andesites.

Physical Properties

Name: Analcite (Analcime)
Group: Zeolite
Formula: NaAlSi2O6+H2O
Crystal System: Cubic (also tet., orth., mon.)
Color: Colorless, white, grey, yellowish, pink, greenish
Opacity: Transparent to translucent
Luster: Vitreous
Streak: Colorless
SGLow: 2.22
SGHigh: 2.27
HardnessLow: 5
HardnessHigh: 5.5
Cleavage: 1
Direction: {001} in traces
Habit: Crystals usually well-formed trapezohedrons or modified cubes; massive granular
Fracture: Subconchoidal; brittle
Other: Occurs in alkali igneous rocks. In cavities and veins within lavas, particularly basalts
Comments: Series with pollucite

Optical Properties

Name: Analcite (Analcime)
Group: Zeolite
Formula: NaAlSi2O6+H2O
Crystal System: Isometric, cubic (also tet., orth., mon.)
Color: Colorless
Form: Euhedral, trapezohedrons as octagonal to rounded crystlas. Irregular masses in groundmass of volcanic rocks.
Relief: Moderate, n<balsam
Birefringence: Isometric, sometimes shows weak birefringence
2V:
Nalpha or Nord.: 1.487
NBeta or Nextr.:
NGamma:
Optical Sign:
Orientation:
Pleochroism:
Twinning:
Cleavage: Imperfect cubic cleavage, two sets at right angles
Extinction: Isotropic
Alteration:
Features: Octagonal crystals, isotropism and RI distinctive
Occurrence: Occurs in alkali igneous rocks. In cavities and veins within lavas, particularly basalts

Physical Properties

Name: Andalusite (Chiastolite)
Group: Andalusite
Formula: Al2SiO5
Crystal System: Orthorhombic
Color: Pink, reddish brown, rose-red, whitish; greyish, yellowish, violet, greenish, blue
Opacity: Transparent to nearly opaque
Luster: Vitreous to subvitreous
Streak: Colorless
SGLow: 3.13
SGHigh: 3.16
HardnessLow: 6.5
HardnessHigh: 7.5
Cleavage: 2
Direction: {110} distinct, {100} indistinct
Habit: Crystals prismatic, nearly square in cross section; massive, compact
Fracture: Subconchoidal to uneven; brittle
Other: Widespread within gneisses, schists, slates, hornfelses and other metamorphic rocks
Comments: Trimorphous with kyanite and sillimanite, series with kanonaite

Optical Properties

Name: Andalusite (Chiastolite)
Group: Sillaminite
Formula: Al2SiO5
Crystal System: Orthorhombic
Color: Colorless, rarely reddish
Form: Euhedral square shaped crystals and columnar aggregates
Relief: High, n>balsam
Birefringence: Weak, 0.007-0.011
2V: 84"
Nalpha or Nord.: 1.629-1.640
NBeta or Nextr.: 1.633-1.644
NGamma: 1.639-1.647
Optical Sign: Biaxial negative
Orientation: Length fast, r>v weak
Pleochroism: Colored varieties are plaochroic from rose red-pale green
Twinning: Penetration twins common
Cleavage: Distinct parallel to {110}. Cross sections show two directions at right angles.
Extinction: Parallel in columnar sections, symmetrical in cross sections
Alteration: Alters to sillaminite
Features: Distinguished from sillimanite by cleavage at right angles, pleochroism, weak birefringence
Occurrence: Widespread within gneisses, schists, slates, hornfelses and other metamorphic rocks

Physical Properties

Name: Andesine

Group: Feldspar

Formula: (Na,Ca)Al(Al,Si)Si2O8

Crystal System: Triclinic

Color: Colorless, white, grey

Opacity: Transparent to translucent

Luster: Vitreous

Streak: White

SGLow: 2.66

SGHigh: 2.68

HardnessLow: 6

HardnessHigh: 6.5

Cleavage: 3

Direction: {001} perfect, {010} nearly perfect; {110} imperfect

Habit: Crystals tabular, uncommon; massive, cleavable, granular

Fracture: Conchoidal to uneven; brittle

Other: Most abundant in felsic igneous rocks. Widespread occurrence

Comments: Intermediate in plagioclase series

Optical Properties

Name: Andesine

Group: Feldspar

Formula: (Na,Ca)Al(Al,Si)Si2O8 An30-50%

Crystal System: Triclinic

Color: Colorless

Form: Anhedral to euhedral triclinic pinacoids

Relief: Low, n always>balsam

Birefringence: Weak, 0.008

2V: 76"-90"

Nalpha or Nord.: 1.543-1.544

NBeta or Nextr.: 1.548-1.558

NGamma: 1.551-1.562

Optical Sign: Biaxial positive or negative

Orientation: r<v

Pleochroism:

Twinning: According to albite, carlsbad and pericline laws, as in albite

Cleavage: Three, perfect, {001}, distinct {010}, imperfect {110} and {11_0}

Extinction: Oblique. In albite twins 13"-27.5". Cleavage sections on {001} 0"- minus 7", on{010} 0"- minus 16"

Alteration: Alters to clay

Features: Maximum extinction angles of albite twins distinctive and RI.

Occurrence: Most abundant in felsic igneous rocks, such as diorites and andesites. Ocurs in some metamorphic rocks. An50 marks boundary between silicic (acid) and calcic (basic) rock types. Widespread occurrence

Physical Properties

Name: Andradite
Group: Garnet
Formula: Ca3Fe2(SiO4)3
Crystal System: Cubic
Color: Shades of yellowish green, green, greenish brown, brown, reddish brown, greyish black
Opacity: Transparent to nearly opaque
Luster: Vitreous to resinous
Streak: Colorless
SGLow: 3.7
SGHigh: 4.1
HardnessLow: 6.5
HardnessHigh: 7
Cleavage: 0
Direction: None
Habit: Crystals usually dodecahedrons or trapezohedrons; massive; granular
Fracture: Uneven to conchoidal; brittle
Other: Occurs within contact metamorphic zones
Comments: Series with grossular and with schorlomite

Optical Properties

Name: Andradite
Group: Garnet
Formula: Ca3Fe2(SiO4)3
Crystal System: Isometric, cubic
Color: Colorless to pale red, pale brown to brown, greenish
Form: Euhedral dodecahedrons in six sided trapezohedrons in eigth sided cross sections. Plygonal grains and aggregates.
Relief: V.high, n>balsam
Birefringence: Isotropic, but may show weak birefringence
2V:
Nalpha or Nord.: 1.857-1.887
NBeta or Nextr.:
NGamma:
Optical Sign:
Orientation:
Pleochroism:
Twinning:
Cleavage: Parting parallel to{110}, irregular fractures
Extinction:
Alteration:
Features: Similar to spinel which is octohedral. Determination of RI will differentiate garnets.
Occurrence: Occurs within contact metamorphic zones

Physical Properties

Name: Anhydrite

Group: Evaporites

Formula: CaSO4

Crystal System: Orthorhombic

Color: White

Opacity: Translucent to transparent

Luster: Non metallic

Streak: White to greyish

SGLow: 2.9

SGHigh: 3.0

HardnessLow: 3

HardnessHigh: 3.5

Cleavage: 3

Direction: {010} perfect, {100} very good, {001} good

Habit: Massive aggragates of parallel, radiating fibers. Tabular

Fracture: Uneven

Other: Found within sedimentary evaporite deposits. Main constituent of anhydrite rock

Comments: Major evaporite mineral

Optical Properties

Name: Anhydrite

Group: Evaporites

Formula: CaSO4

Crystal System: Orthorhombic

Color: Colorless

Form: Anhedral to subhedral aggregates of fine to medium grains, occasionally elongate

Relief: Moderate, n>balsam

Birefringence: Strong, 0.044

2V: 42"

Nalpha or Nord.: 1.570

NBeta or Nextr.: 1.576

NGamma: 1.614

Optical Sign: Biaxial positive

Orientation:

Pleochroism:

Twinning: Polysynthetic with {101} Forms angles 42" and 48" to cleavage traces. Sometimes two sets of twin lamellae intersection at 83.5" and 96.5"

Cleavage: Three directions at right angles. Parallel to {100}, {010} and {001}. May show parting parallel to {101) due to twinning

Extinction: Parallel to cleavage

Alteration: Often altered to gypsum

Features: Strong birefringence, rectangular pseudo cubic cleavage distinguishes from gypsum

Occurrence: Found within sedimentary evaporite deposits. Main constituent of anhydrite rock

Physical Properties

Name: Anorthite
Group: Feldspar
Formula: CaAl2Si2O8
Crystal System: Triclinic
Color: Colorless, white, greyish, reddish
Opacity: Transparent to translucent
Luster: Vitreous
Streak: White
SGLow: 2.74
SGHigh: 2.76
HardnessLow: 6
HardnessHigh: 6.5
Cleavage: 3
Direction: {001} perfect, {010} nearly perfect, {110} imperfect
Habit: Crystals usually short prismatic; massive, cleavable
Fracture: Conchoidal to uneven; brittle
Other: Quite rare compared to other plagioclase feldspars, occurs in some lavas and contact metamorphic rocks.
Comments:

Optical Properties

Name: Anorthite
Group: Feldspar
Formula: CaAl2Si2O8 An90-100%
Crystal System: Triclinic
Color: Colorless
Form: Anhedral to subhedral plates and lathes
Relief: Fair, n>balsam
Birefringence: Weak, 0.012-0.013
2V: 77"-79"
Nalpha or Nord.: 1.573-1.577
NBeta or Nextr.: 1.579-1.585
NGamma: 1.585-1.590
Optical Sign: Biaxial negative
Orientation: r>v
Pleochroism:
Twinning: According to albite, carlsbad and pericline laws, as in albite
Cleavage: Four, perfect {001}, distinct {010}, imperfect {110} and {11_0}
Extinction: Oblique, 51"-70" in albite twins. Cleavage sections are minus 32" - minus 40" on {001} and minus 37" on {010}
Alteration: Alters to clay
Features: Maximum extinction angles in albite twins and RI are distinctive
Occurrence: Quite rare compared to other plagioclase feldspars. occurs in some lavas and contact metamorphic rocks.

Physical Properties

Name: Anthophyllite

Group: Amphibole

Formula: $(Mg,Fe)_7Si_8O_{22}(OH)_2$

Crystal System: Orthorhombic

Color: White, grey, greenish, brownish green, clove-brown, dark brown

Opacity: Transparent to nearly opaque

Luster: Vitreous to silky

Streak: Colorless or greyish

SGLow: 2.85

SGHigh: 3.57

HardnessLow: 5.5

HardnessHigh: 6

Cleavage: 3

Direction: {110} perfect, {010} and {100} imperfect

Habit: Crystals prismatic, rare; massive, fibrous, or lamellar

Fracture:

Other: Occurs in metamorphic rocks

Comments: Series with magnesio-anthophyllite and ferro-anthophyllite

Optical Properties

Name: Anthophyllite

Group: Amphibole

Formula: $(Mg,Fe)_7(OH)_2(Si_4O_{11})_2$

Crystal System: Orthgorhombic

Color: Colorless to pale

Form: Long prismatic amphibole habit and columnar to fibrous aggregates

Relief: High, n>balsam

Birefringence: Moderate, 0.016-0.025

2V: 70"-90"

Nalpha or Nord.: 1.598-1.652

NBeta or Nextr.: 1.615-1.662

NGamma: 1.623-1.676

Optical Sign: Biaxial positive

Orientation: Length slow, r>v or r<v

Pleochroism: Colored varieties are pleochroic

Twinning:

Cleavage: In two directions {110} at 54" and 126"

Extinction: Parallel in longitudinal sections, symmetrical in cross sections

Alteration: Alters to talc

Features: Parallel extinction distinguishes from tremolite/actinolite

Occurrence: Characteristic of metamorphic rocks, particularly schists. A secondary mineral in some metamorphosed dunites and peridotites. Principle constituent of asbestos.

Physical Properties

Name: Antigorite
Group: Serpentine
Formula: (MgFe)3Si2O5(OH)4
Crystal System: Monoclinic
Color: Green, green-blue, white,
Opacity: Opaque
Luster: Greasy, waxy, silky
Streak: White
SGLow: 2.52
SGHigh: 2.6
HardnessLow: 2.5
HardnessHigh: 3.5
Cleavage: 1
Direction: {001} perfect
Habit: Usually massive
Fracture: Conchoidal
Other: Main serpentine group mineral. Comprises serpentine rock. Alteration product of pyroxenes and olivine. Metamorphic mineral
Comments: Main constituent of serpentine rock

Optical Properties

Name: Antigorite
Group: Serpentine
Formula: (MgFe)3Si2O5(OH)4
Crystal System: Monoclinic
Color: Colorless to pale green
Form: Anhedral crystals or fibrous and flaky lamellar aggregates
Relief: Low, n>balsam
Birefringence: Weak, 0.007-0.009
2V: 20"-90"
Nalpha or Nord.: 1.555-1.564
NBeta or Nextr.: 1.562-1.573
NGamma: 1.562-1.573
Optical Sign: Biaxial negative
Orientation: Length slow
Pleochroism:
Twinning:
Cleavage: One perfect {001}
Extinction: Parallel
Alteration:
Features: Similar to chrysotile but aggregate structure distinctive
Occurrence: Main serpentine group mineral. Comprises serpentine rock. Alteration product of pyroxenes and olivine. Metamorphic mineral

Physical Properties

Name: Apatite

Group: Apatite

Formula: ca5(PO4)3(F,Cl,OH)

Crystal System: Hexagonal

Color: Green, brown, red colorless yellow, pink, white, violet

Opacity: Transparent to translucent

Luster: Greasy, vitreous

Streak: White

SGLow: 3.1

SGHigh: 3.2

HardnessLow: 5

HardnessHigh: 5

Cleavage: Poor, 1 direction

Direction: Crosswise

Habit: Short or long prismatic, tabular in igneous rocks. Oolitic, earthy, stalactitic, coarse granular in sedimentary rocks.

Fracture: Uneven, conchoidal

Other: Widely distributed in igneous rocks as an accessory. In some metamorphic rocks, hornfels. In sedimentary phosphoritic rocks.

Comments: Most common phosphate mineral.

Optical Properties

Name: Apatite

Group: Apatite

Formula: Ca5(PO4)3F

Crystal System: Hexagonal

Color: Colorless

Form: Minute, hexagonal crystals

Relief: Moderate, n>balsam

Birefringence: Weak, 0.003-0.004

2V:

Nalpha or Nord.: 1.630-1.651

NBeta or Nextr.: 1.633-1.655

NGamma:

Optical Sign: Uniaxial negative

Orientation: Length fast, tabular crystals are length slow

Pleochroism:

Twinning:

Cleavage: Two, imperfect basal {0001}

Extinction: Parallel

Alteration:

Features: Hexagonal crystals and birefringence

Occurrence: Widespread accessory mineral in igneous rocks.. In pegmatites, veins, metamorphic limestones and sedimentary beds.

Physical Properties

Name: Aragonite
Group: Aragonite
Formula: CaCO3
Crystal System: Orthorhombic
Color: Colorless, white, yellowish, grey, green, bluish, lavender, reddish, brown
Opacity: Transparent to translucent
Luster: Vitreous to resinous
Streak: White
SGLow: 2.95
SGHigh: 2.95
HardnessLow: 3.5
HardnessHigh: 4
Cleavage: 3
Direction: {010} distinct, {110} and {011} indistinct
Habit: Crystals acicular or chisel-shaped, elongated, thick tabular or pyramidal; frequently twinned as sixlings; columnar, fibrous
Fracture: Subconchoidal; brittle
Other: A secondary mineral in cavities and seams within basalts and andesites, limestones and sandstones.
Comments: Trimorphous with calcite and vaterite; often fluoresces greenish white, green, yellowish, pink or bluish

Optical Properties

Name: Aragonite
Group: Aragonite
Formula: CaCO3
Crystal System: Orthorhombic pseudo hexagonal
Color: Colorless
Form: Columnar or fibrous, hexagonal cross sections
Relief: Low when columns and fibres are parallel to polariser, high when at right angles
Birefringence: Extreme, 0.156
2V: 18"
Nalpha or Nord.: 1.530
NBeta or Nextr.: 1.682
NGamma: 1.686
Optical Sign: Biaxial negative
Orientation:
Pleochroism:
Twinning: Parallel to crystals and columns
Cleavage: Three {010} distinct, {110} and {011} indistinct, imperfect parallel to length of crystals {010}
Extinction: Parallel to crystals or columns
Alteration: Alters to calcite
Features: Resembles calcite, no rhombohedral cleavage, biaxial.
Occurrence: A secondary mineral in cavities and seams within basalts and andesites, limestones and sandstones. Alters easily to calcite.

Physical Properties

Name: Arfvedsonite
Group: Amphibole
Formula: Na3(Fe,Mg)4FeSi8O22(OH)2
Crystal System: Monoclinic
Color: Greenish black, black
Opacity: Nearly opaque
Luster: Vitreous
Streak: Dark bluish grey
SGLow: 3.37
SGHigh: 3.50
HardnessLow: 5
HardnessHigh: 6
Cleavage: 2
Direction: {110} perfect; {010} parting
Habit: Crystals short to long prismatic, often tabular; prismatic aggregates
Fracture: Uneven; brittle
Other:
Comments: Series with magnesio-arfvedsonite

Optical Properties

Name: Arfvedsonite
Group: Amphibole
Formula: Na3(Fe,Mg)4FeSi8O22(OH)2
Crystal System: Monoclinic
Color: Dark bluish grey
Form: Crystals short to long prismatic, often tabular; prismatic aggregates
Relief:
Birefringence:
2V:
Nalpha or Nord.:
NBeta or Nextr.:
NGamma:
Optical Sign:
Orientation:
Pleochroism:
Twinning:
Cleavage: Two {110} perfect; {010} parting
Extinction:
Alteration:
Features: Series with magnesio-arfvedsonite
Occurrence:

Physical Properties

Name: Augite
Group: Pyroxene
Formula: (Ca,Mg,Fe)2Si2O6
Crystal System: Monoclinic
Color: Pale brown to dark brown or purplish brown, greenish, black
Opacity: Translucent to nearly opaque
Luster: Vitreous to dull
Streak: Greyish green
SGLow: 3.23
SGHigh: 3.52
HardnessLow: 5.5
HardnessHigh: 6
Cleavage: 3
Direction: {110} good; {100} and {010} parting
Habit: Crystals short prismatic; massive, granular, rarely fibrous
Fracture: Uneven to conchoidal; brittle
Other: Characteristic of mafic and ultramafic igneous and metamorphic rocks
Comments:

Optical Properties

Name: Augite
Group: Pyroxene
Formula: (Ca,Mg,Fe)2Si2O6
Crystal System: Monoclinic
Color: Colorless, neutral, pale green or pale purplish brown
Form: Subhedral prismatic crystals with four or eight sided cross sections
Relief: High, n>balsam
Birefringence: Moderate, 0.021-0.025
2V: 58"-62"
Nalpha or Nord.: 1.688-1.712
NBeta or Nextr.: 1.701-1.717
NGamma: 1.713-1.737
Optical Sign: Biaxial positive
Orientation: Length fast, r>v
Pleochroism: Uaually absen but may show weak pleochroism on {100} sections
Twinning: Simple twins alon {100} are common. Polysynthetic twinning along {001} sometimes. Polysynthetic twinnning gives herringbone structure
Cleavage: Three, one direction along longitudinal sections, two directions in cross sections at 87" and 93"
Extinction: Oblique 36"-45" in longitudinal sections and parallel in cross sections. Wavy extinction gives hourglass structure
Alteration: Hornblende during late magmatic stage, secondary actinolite/tremolite by hydrothermal alteration
Features: Similar to diopside, occurrence will differentiate.
Occurrence: Characteristic of mafic and ultramafic igneous and metamorphic rocks, including basalts, gabbros, peridotites, gniesses and granulites

Physical Properties

Name: Axinite

Group: Axinite

Formula: $(Ca, Mn, Fe, Mg)_3Al_2BSi_4O_{15}(OH)$

Crystal System: Triclinic

Color: Yellow, greenish, grey, black, brown

Opacity: Transparent to translucent

Luster: Vitreous

Streak: Colorless

SGLow: 3.3

SGHigh: 3.4

HardnessLow: 6.5

HardnessHigh: 7.0

Cleavage: Good, one direction

Direction:

Habit: Flattened, tabular, striated crystals. Lamellar, granular and bladed crystals

Fracture: Uneven to conchoidal

Other: Occurs in contact metamorphhic hornfels with diopside and andradite and with quartz and calcite in high temperature hydrothermal veins.

Comments:

Optical Properties

Name: Axinite

Group: Garnet

Formula: $H(Fe, Mn)Ca_2Al_2B(SiO_4)_4$

Crystal System: Triclinic

Color: Colorless to pale violet

Form: Elongated anhedral to subhedral prisms with acute- angled sections. Bladed to wedge shaped.

Relief: High, n>balsam

Birefringence: Weak, 0.010-0.012

2V: 70"-75"

Nalpha or Nord.: 1.678-1.684

NBeta or Nextr.: 1.685-1.692

NGamma: 1.688-1.696

Optical Sign: Biaxial negative

Orientation: r<v or r>v

Pleochroism: Thick sections may be pleochroic

Twinning:

Cleavage: Imperfect in several directions

Extinction: Oblique to cleavage and outline.

Alteration:

Features: Bladed to wedge shaped crystal form, low birefringence and biaxial interference figure.

Occurrence: Occurs chiefly in calcareous rocks in contact metamorphic zones. Less common in gra.nites and granite pegmatites

Physical Properties

Name: Baryte (Barite)
Group: Barite
Formula: $BaSO_4$
Crystal System: Orthorhombic
Color: Colorless, white to greyish, yellowish to brown, bluish, greenish, reddish
Opacity: Transparent to subtranslucent
Luster: Vitreous to resinous
Streak: White
SGLow: 4.50
SGHigh: 4.50
HardnessLow: 3
HardnessHigh: 3.5
Cleavage: 3
Direction: {001} perfect, {210} distinct, {010} imperfect
Habit: Crystals thin to thick tabular; short to long prismatic equant; aggregates, stalactitic, columnar, fibrous, earthy
Fracture: Uneven; brittle
Other: Widespread vein mineral, with quartz and calcite and in some limestones and sandstones
Comments: Sometimes fluoresces or phosphorescent

Optical Properties

Name: Baryte (Barite)
Group: Barite
Formula: $BaSO_4$
Crystal System: Orthorhombic
Color: Colorless
Form: Granular aggregates
Relief: Fairly high, n>balsam
Birefringence: Weak, 0.012
2V: 36"-37.5"
Nalpha or Nord.: 1.636
NBeta or Nextr.: 1.637
NGamma: 1.648
Optical Sign: Biaxial positive
Orientation:
Pleochroism:
Twinning: Polysynthetic with {110} sometimes
Cleavage: Three directions, parallel to {001} and {110} and at 90" and 78"
Extinction: Parallel to main cleavage, symmetrical to {001}
Alteration:
Features: Sometimes fluoresces or phosphorescent. Similar to celestite, chemical tests may be necessary.
Occurrence: Widespread vein mineral, with quartz and calcite and in some limestones and sandstones

Physical Properties

Name: Beryl
Group: Beryl
Formula: Be3Al2Si6O18
Crystal System: Hexagonal
Color: Colorless, white, light green to green, yellow, pink to pinkish orange, red, blue
Opacity: Transparent to translucent
Luster: Vitreous
Streak: Colorless
SGLow: 2.6
SGHigh: 2.9
HardnessLow: 7.5
HardnessHigh: 8
Cleavage: 1
Direction: {0001} indistinct
Habit: Crystals short to long prismatic; often etched
Fracture: Conchoidal to uneven; brittle
Other: Occurs within granite pegmatites, mica schists and in veins in limestones
Comments:

Optical Properties

Name: Beryl
Group: Beryl
Formula: Be3Al2Si6O18
Crystal System: Hexagonal
Color: Colorless
Form: Usually large prismatic crystals and hexagonal cross sections
Relief: Moderate, n>balsam
Birefringence: Weak, 0.004-0.008
2V:
Nalpha or Nord.: 1.564-1.590
NBeta or Nextr.: 1.568-1.598
NGamma:
Optical Sign: Uniaxial negative
Orientation: Length fast
Pleochroism: Colored gem varieties are pleochroic
Twinning:
Cleavage: One, imperfect to {0001}, not usually visible
Extinction: Parallel in longitudinal sections. Basal sections are dark
Alteration: Alters to kaolinite
Features: Hexagonal crystals with low birefringence. Similar to quartz, but ptically negative and length fast. Similar to apatite but lower RI.
Occurrence: Occurs within granite pegmatites, mica schists and in veins in limestones

Physical Properties

Name: Biotite

Group: Mica

Formula: $K(Mg,Fe)_3(Al,Fe)Si_3O_{10}(OH,F)_2$

Crystal System: Monoclinic

Color: Black, dark brown, reddish brown, green, rarely white

Opacity: Transparent to opaque

Luster: Splendent, submetallic, vitreous

Streak: Colorless

SGLow: 2.7

SGHigh: 3.4

HardnessLow: 2.5

HardnessHigh: 3

Cleavage: 1

Direction: {001} perfect

Habit: Crystals tabular or short prismatic, pseudohexagonal outline; disseminated, massive aggregates

Fracture: Thin laminae flexible to brittle

Other: Widespread within felsic igneous rocks, gneisses, schists and low grade metamorphic rocks

Comments: Series with phlogopite

Optical Properties

Name: Biotite

Group: Mica

Formula: $K(Mg,Fe)_3(Al,Fe)Si_3O_{10}(OH,F)_2$

Crystal System: Monoclinic

Color: Brown, yellowish-brown, reddish-drown, olive green, green

Form: Euhedral thick tabular six sided crystals and elongated plates

Relief: Fair, n>balsam

Birefringence: Strong, 0.033-0.059

2V: 0"-25"

Nalpha or Nord.: 1.541-1.579

NBeta or Nextr.: 1.574-1.638

NGamma: 1.574-1.638

Optical Sign: Biaxial negative

Orientation: Length slow, r>v or r<v weak

Pleochroism: Strong pleochroism from brown - yellowish brown or reddish brown and olive green - green

Twinning: According to mica law {110}

Cleavage: One perfect {001}

Extinction: Parallel, upto 3"

Alteration: Alters to chlorite

Features: Similar to phlogopite but darker and stronger pleochroism

Occurrence: Widespread within felsic igneous rocks, gneisses, schists and low grade metamorphic rocks and contact metamorphic zones

Physical Properties

Name: Boehmite (Bauxite Group)

Group: Bauxite

Formula: ALO(OH)

Crystal System: Orthorhombic

Color: White, brown

Opacity: Transparent to opaque

Luster: Dull

Streak: White

SGLow: 3.070

SGHigh: 3.070

HardnessLow: 3

HardnessHigh: 3

Cleavage: 1

Direction: {010} very good

Habit: Crystals massive to microscopic, tabular; pisolitic aggregates, disseminated

Fracture: Earthy

Other: Occurs as a surface weathering product from tropical weathering

Comments: Dimorphous with diaspore

Optical Properties

Name: Boehmite (Bauxite Group)

Group: Bauxite

Formula: ALO(OH)

Crystal System: Orthorhombic

Color: Colorless

Form: Minute, tabular crystals

Relief: Moderate

Birefringence: Moderate, 0.013

2V: Moderate

Nalpha or Nord.: 1.638

NBeta or Nextr.: 1.645

NGamma: 1.651

Optical Sign: Biaxial neg?

Orientation:

Pleochroism:

Twinning:

Cleavage: One direction parallel to {010}

Extinction:

Alteration: Dimorphous with diaspore

Features: Birefringence, crystals, closely resembles gibbsite

Occurrence: Occurs as a surface weathering product from tropical weathering in bauxite deposits.

Physical Properties

Name: Borax
Group: Borax
Formula: Na2B4O5(OH)4+8H2O
Crystal System: Monoclinic
Color: Colorless, white, greyish, greenish, bluish
Opacity: Transparent to opaque
Luster: Vitreous, sometimes earthy
Streak:
SGLow: 1.715
SGHigh: 1.715
HardnessLow: 2
HardnessHigh: 2.5
Cleavage: 2
Direction: {100} perfect, {110} imperfect
Habit: Crystals short prismatic, somewhat tabular
Fracture: Conchoidal; brittle
Other: Associated with evaporite minerals in evaporite deposits
Comments: Soluble in water

Optical Properties

Name: Borax
Group: Borax
Formula: Na2B4O5(OH)4+8H2O
Crystal System: Monoclinic
Color: White to greyish, greenish, bluish
Form: Stubby, prismatic crystals
Relief: Low to moderate, n<balsam
Birefringence: Moderate to strong, 0.025
2V: 40"
Nalpha or Nord.: 1.447
NBeta or Nextr.: 1.469
NGamma: 1.472
Optical Sign: Biaxial negative
Orientation: Strong crossed dispersion, r>v, optic plane at right angles to {010}
Pleochroism:
Twinning:
Cleavage: Two {100} perfect, {110} fair, {010} traces
Extinction:
Alteration:
Features: Soluble in water
Occurrence: Associated with evaporite minerals in evaporite deposits

Physical Properties

Name: Bronzite
Group: Pyroxene
Formula: MgFeSiO3
Crystal System: Orthorhombic
Color: Brown
Opacity: Transparent to translucent
Luster: Non Metallic
Streak: Colorless
SGLow: 3.3
SGHigh: 3.3
HardnessLow: 5
HardnessHigh: 6
Cleavage: 2
Direction: {210} good, {100} and {010} parting
Habit: Crystals prismatic; massive, lamellar, fibrous
Fracture: Uneven; brittle
Other: Characteristic of ultramafic igneous rocks and metamorphic serpentinites
Comments: A ferroan variety of enstatite

Optical Properties

Name: Bronzite (Enstatite)
Group: Pyroxene
Formula: MgFeSiO3
Crystal System: Orthorhombic
Color: Colorless to neutral
Form: Crystals prismatic; massive, lamellar, fibrous
Relief: High, n>balsam
Birefringence: Weak, 0.008-0.009
2V: 58"-80"
Nalpha or Nord.: 1.650-1.665
NBeta or Nextr.: 1.653-1.670
NGamma: 1.658-1.674
Optical Sign: Biaxial positive
Orientation: Length slow, r<v weak
Pleochroism: Weak pleochroism
Twinning: Rare
Cleavage: Two {210} good, {100} and {010} parting
Extinction: Parallel
Alteration: Alters to antigorite
Features: Similar to hypersthene, but weak or no pleochroism. Parallel extinction distinguishes from clino pyroxenes.
Occurrence: Characteristic of ultramafic igneous rocks and metamorphic serpentinites

Physical Properties

Name: Brucite
Group: Bauxite
Formula: Mg(OH)2
Crystal System: Trigonal
Color: White, pale green, grey, grey-blue, blue
Opacity: Transparent
Luster: Pearly, waxy, vitreous
Streak: White
SGLow: 2.39
SGHigh: 2.39
HardnessLow: 2.5
HardnessHigh: 2.5
Cleavage: 1
Direction: {0001} perfect
Habit: Crystals broad tabular; acicular; foliated massive, scaly, fine granular
Fracture: Plates separable and flexible; sectile
Other: Occurs within metamorphic limestones as an alteration of periclase
Comments:

Optical Properties

Name: Brucite
Group: Bauxite
Formula: Mg(OH)2
Crystal System: Hexagonal
Color: Colorless
Form: Plates or scaley aggregates that appear fibrous
Relief: Fair, n>balsam
Birefringence: Moderate 0.019, anomolous reddish brown hue
2V:
Nalpha or Nord.: 1.566
NBeta or Nextr.: 1.585
NGamma:
Optical Sign: Uniaxioal positive
Orientation: Length fast
Pleochroism:
Twinning:
Cleavage: Perfect one direction, {0001}
Extinction: Parallel
Alteration: Often altered to hydro magnesite
Features: Cleavage, anomolous interference colors
Occurrence: Occurs within metamorphic limestones as an alteration of periclase

Physical Properties

Name: Bytownite

Group: Feldspar

Formula: (Ca,Na)(Si,Al)4O8

Crystal System: Triclinic

Color: Colorless, white, grey

Opacity: Transparent to translucent

Luster: Vitreous

Streak: White

SGLow: 2.72

SGHigh: 2.74

HardnessLow: 6

HardnessHigh: 6.5

Cleavage: 3

Direction: {001} perfect, {010} nearly perfect, {110} imperfect

Habit: Crystals tabular; massive, granular

Fracture: Conchoidal to uneven; brittle

Other: Most abundant in felsic igneous rocks. Widespread occurrence

Comments: Plagioclase series

Optical Properties

Name: Bytownite

Group: Feldspar

Formula: (Ca,Na)(Si,Al)4O8 An70-90%

Crystal System: Triclinic

Color: Colorless

Form: Anhedral to subhedral triclininc pinacoids

Relief: Moderate, n>balsam

Birefringence: 2.72

2V: 79"-88"

Nalpha or Nord.: 1.564-1.573

NBeta or Nextr.: 1.569-1.579

NGamma: 1.573-1.585

Optical Sign: Biaxial negative

Orientation: r>v

Pleochroism:

Twinning: According to albite, carlsbad and pericline laws, as in albite

Cleavage: Four perfect {001}, distinct {010}, imperfect {110} and {11_0}

Extinction: Oblique 39"-51" in albite twins. On cleavage sections -minus16" - minus 32" on {001} and minus 29" - minus 36" on {010}

Alteration: Alters to clay

Features: Extinction angles on albite twins and RI are distinctive

Occurrence: Mostly occurs in mafic igneous rocks., such as gabbros, basalts and anorthosites, but is somewhat rare. Widespread occurrence

Physical Properties

Name: Calcite
Group: Calcite
Formula: CaCO3
Crystal System: Trigonal
Color: Colorless or white, grey, yellow, brown, red, green, blue, black
Opacity: Transparent to translucent
Luster: Vitreous to pearly, dull
Streak: White to greyish
SGLow: 2.710
SGHigh: 2.710
HardnessLow: 3
HardnessHigh: 3
Cleavage: 3
Direction: {10-11} perfect, {01-12} and {0001} parting
Habit: Crystals varied, scalenohedrons and rhombohedrons common; massive, granular, stalactitic, fibrous
Fracture: Conchoidal; brittle
Other: The major constituent of sedimentary and metamorphic limestones. The most common vein mineral, with quartz
Comments: Trimorphous with aragonite and vaterite; series with rhodochrosite

Optical Properties

Name: Calcite
Group: Calcite
Formula: CaCO3
Crystal System: Hexagonal
Color: Colorless
Form: Anhedral fine to coarse aggragates, oolitic, spherulitic, organic structure
Relief: High in long direction, low in short direction
Birefringence: Extreme, 0.172
2V:
Nalpha or Nord.: 1.486
NBeta or Nextr.: 1.658
NGamma:
Optical Sign: Uniaxial negative
Orientation:
Pleochroism:
Twinning: Polysynthetic twinning on {011_2}
Cleavage: Three rhombohedral {10-11} perfect, {01-12} and {0001} parting
Extinction: Symmetrical to cleavage
Alteration: Often replaced by quartz
Features: Dolomite, siderite and magnesite are very similar. Dolomite is usually subhedral to euhedral. Siderite usually has iron stainings. Staining may be required to distinguish from magnesite.
Occurrence: The major constituent of sedimentary and metamorphic limestones. The most common vein mineral, with quartz

Physical Properties

Name: Cancrinite
Group: Feldspathoid
Formula: Na6Ca2Al6Si6O24(CO3)2
Crystal System: Hexagonal
Color: Colorless, white, yellow, orange, pink to reddish, pale blue to pale bluish grey
Opacity: Transparent to translucent
Luster: Vitreous, pearly or greasy
Streak: Colorless
SGLow: 2.42
SGHigh: 2.51
HardnessLow: 5
HardnessHigh: 6
Cleavage: 2
Direction: {10-10} perfect, {0001} poor
Habit: Crystals prismatic; massive
Fracture: Uneven; brittle
Other: Occurs within alkali igneous rocks
Comments:

Optical Properties

Name: Cancrinite
Group: Feldspathoid
Formula: Na6Ca2Al6Si6O24(CO3)2
Crystal System: Hexagonal
Color: Colorless
Form: Usually anhedral , euhedral crystals rare.
Relief: Fair, n<balsam
Birefringence: Weak, 0.003-0.004
2V:
Nalpha or Nord.: 1.496-1.500
NBeta or Nextr.: 1.507-1.524
NGamma:
Optical Sign: Uniaxial negative
Orientation:
Pleochroism:
Twinning:
Cleavage: Two {10-10} perfect, {0001} poor
Extinction: Parallel to cleavage and outline
Alteration:
Features: Stronger birefringence
Occurrence: Occurs within alkali igneous rocks, particularly nepheline syenites, rare but widespread.

Physical Properties

Name: Carnotite
Group: Carnotite
Formula: K2(UO2)2V2O8+3H2O
Crystal System: Monoclinic
Color: Bright yellow to golden yellow, greenish yellow
Opacity: Transparent to translucent
Luster: Pearly masses dull or earthy
Streak: Yellow
SGLow: 4.70
SGHigh: 4.70
HardnessLow: 2
HardnessHigh: 2
Cleavage: 1
Direction: {001} perfect, micaceous
Habit: Crystals microscopic, rhomboidal or diamond-shaped outline; disseminated, coatings, powdery, masses, crusts, powdery
Fracture: Sectile
Other: A secondary mineral occuring in sedimentary sandstone beds
Comments:

Optical Properties

Name: Carnotite
Group: Carnotite
Formula: K2(UO2)2V2O8+3H2O
Crystal System: Monoclinic
Color: Yellow to nearly colorless(X)
Form: Microcrystalline aggregates
Relief: High
Birefringence: Extreme, 0.20
2V: 40"-50"
Nalpha or Nord.: 1.750
NBeta or Nextr.: 1.925
NGamma: 1.950
Optical Sign: Biaxial negative
Orientation:
Pleochroism: Pleochroic colorless-canary yellow
Twinning:
Cleavage: One perfect {001}
Extinction:
Alteration:
Features: Color, microcrystalline form, extreme birefringence, fluorescent in uv.
Occurrence: A secondary mineral occuring in sedimentary sandstone beds

Physical Properties

Name: Cassiterite
Group: Rutile
Formula: SnO2
Crystal System: Tetragonal
Color: Brown to brownish black, black, colorless, grey, yellowish, greenish, red
Opacity: Transparent to nearly opaque
Luster: Adamantine, vitreous
Streak: White, greyish, brown
SGLow: 6.99
SGHigh: 6.99
HardnessLow: 6
HardnessHigh: 7
Cleavage: 4
Direction: {100} imperfect, {110} indistinct, {111} and {011} parting
Habit: Crystals short prismatic, sometimes slender prismatic or pyramidal; granular, botryoidal or reniform masses
Fracture: Subconchoidal to uneven; brittle
Other: Occurs in granite pegmatites, greisen and high temperature veins
Comments:

Optical Properties

Name: Cassiterite
Group: Rutile
Formula: SnO2
Crystal System: Tetragonal
Color: Colorless, yellowish, reddish, greyish, brown
Form: Subhedral, often in veinlets
Relief: V. high, n>balsam
Birefringence: Extreme, 0.097
2V: 6.99
Nalpha or Nord.: 1.966
NBeta or Nextr.: 2.093
NGamma:
Optical Sign: Uniaxial positive
Orientation:
Pleochroism:
Twinning: Common, {101}
Cleavage: Four, prismatic, parallel to length, {100} imperfect, {110} indistinct, {111} and {011} parting
Extinction: Parallel to cleavage
Alteration:
Features: Color, birefringence, cleavage
Occurrence: Occurs in granite pegmatites, greisen and high temperature veins

Physical Properties

Name: Celestite
Group: Barite
Formula: SrSO4
Crystal System: Orthorhombic
Color: Colorless, white, grey, blue, green, yellow, red, brown
Opacity: Transparent to translucent
Luster: Vitreous
Streak: White
SGLow: 3.97
SGHigh: 3.97
HardnessLow: 3
HardnessHigh: 3.5
Cleavage: 3
Direction: {001} perfect, {210} good, {010} indistinct
Habit: Crystals thin to thick tabular, or equant or pyramidal; nodules, fibrous veinlets; massive granular, lamellar, earthy
Fracture: Uneven; brittle
Other: Occurs in sedimentary limestones
Comments: Sometimes fluoresces, name celestine also accepted

Optical Properties

Name: Celestite
Group: Barite
Formula: SrSO4
Crystal System: Orthorhombic
Color: Colorless
Form: Euhedral to anhedral tabular crystals elongated
Relief: Fair, n>balsam
Birefringence: Weak, 0.009
2V: 51"
Nalpha or Nord.: 1.622
NBeta or Nextr.: 1.624
NGamma: 1.631
Optical Sign: Biaxial positive
Orientation: Length slow
Pleochroism:
Twinning:
Cleavage: Three perfect parallel to {001}, imperfect parallel to {110}
Extinction: Parallel to cleavage and outlines
Alteration:
Features: Sometimes fluoresces, name celestine also accepted. Similar to barite, axial angle is larger. Chhemical tests may be required.
Occurrence: Occurs in sedimentary limestones

Physical Properties

Name: Chabazite
Group: Zeolite
Formula: CaAl2Si4O12+6H2O
Crystal System: Trigonal
Color: Colorless, white, yellowish, pinkish, reddish white, greenish
Opacity: Transparent to translucent
Luster: Vitreous
Streak: Colorless
SGLow: 2.05
SGHigh: 2.16
HardnessLow: 4
HardnessHigh: 5
Cleavage: 1
Direction: {10-11} distinct
Habit: Crystals simple rhombohedrons, resemble cubes, may be complex, tabular
Fracture: Uneven; brittle
Other: Secondary mineral in cavities and veins of igneous rocks, especially basalts
Comments:

Optical Properties

Name: Chabazite
Group: Zeolite
Formula: CaAl2Si4O12+6H2O
Crystal System: Monoclinic
Color: Colorless
Form: Euhedral rhombohedrons
Relief: Moderate, n<balsam
Birefringence: Weak, 0.002-0.010
2V: 0" -32"
Nalpha or Nord.: 1.478-1.485
NBeta or Nextr.:
NGamma: 1.480-1.490
Optical Sign: Biaxial positive
Orientation:
Pleochroism:
Twinning:
Cleavage: One {10-11} distinct, imperfect rhombohedron, almost rectangular
Extinction: Symmetrical to cleavage and outline
Alteration:
Features: Rhombohedral cleavage distinctive
Occurrence: Secondary mineral in cavities and veins of igneous rocks, especially basalts

Physical Properties

Name: Chalcedony
Group: Quartz
Formula: SiO2
Crystal System: Cryptocrystalline
Color: Grey, white, colorless
Opacity: Transparent to translucent
Luster: Vitreous
Streak: Colorless
SGLow: 2.7
SGHigh: 2.7
HardnessLow: 7
HardnessHigh: 7
Cleavage: None
Direction: None
Habit: Massive, often in veins
Fracture: Subconchoidal to Conchoidal
Other: A secondary mineral within veins and cavities in igneous rocks. As nodules and bands in limestones, replacing calcareous fossils. In sedimentary cherts and jaspers
Comments: Cryptocrystalline variety of quartz

Optical Properties

Name: Chalcedony
Group: Quartz
Formula: SiO2
Crystal System: Cryptocrystalline
Color: Colorless to pale brown
Form: Massive, often in veins, radial or mosaic structure, fibrous, spherulitic aggregates
Relief: Low, n>balsam
Birefringence: Weak, first order grey
2V:
Nalpha or Nord.:
NBeta or Nextr.:
NGamma:
Optical Sign:
Orientation:
Pleochroism:
Twinning:
Cleavage:
Extinction: Parallel to fibers when fibrous
Alteration:
Features: Cryptocrystalline variety of quartz
Occurrence: A secondary mineral within veins and cavities in igneous rocks. As nodules and bands in limestones, replacing calcareous fossils. In sedimentary cherts and jaspers

Physical Properties

Name: Chamosite

Group: Chlorite

Formula: (Fe,Mg)5Al(Si3Al)O10(OH,O)8

Crystal System: Monoclinic

Color: Green, greenish grey to black

Opacity: Transparent to translucent

Luster:

Streak: Colorless

SGLow: 3.12

SGHigh: 3.12

HardnessLow: 3

HardnessHigh: 3

Cleavage: 1

Direction: {001} perfect

Habit: Massive; compact or oölitic

Fracture: Earthy, thin flexible flakes

Other: Characteristic of oolitic sedimentary iron ores

Comments: Dimorphous with orthochamosite; series with clinochlore, H=approximate

Optical Properties

Name: Chamosite

Group: Chlorite

Formula: (Fe,Mg)5Al(Si3Al)O10(OH,O)8

Crystal System: Monoclinic

Color: Almost colorless to green, greenish-grey, grey, pale brown.

Form: Usually oolitic, pseudoshpherulitic. Sometimes thick tabular crystals and massive aggregates

Relief: Moderate, n>balsam

Birefringence: None to weak, 0.007-0.008

2V: Small

Nalpha or Nord.: 1.635

NBeta or Nextr.:

NGamma:

Optical Sign: Biaxial negative

Orientation: Length slow

Pleochroism: Weak pleochroism in some sections

Twinning:

Cleavage: One direction distinct

Extinction:

Alteration:

Features: Occurrence distinctive. Similar to oolitic collophane, but in tabular crystals.

Occurrence: Characteristic of oolitic sedimentary iron ores. Of marine origin, associated with siderite, calcite, collophane and detrital minerals.

Physical Properties

Name: Chlorite
Group: Chlorite
Formula: $(Mg,Fe)_5Al(Si_3Al)O_{10}(OH)_8$
Crystal System: Monoclinic
Color: Yellowish, pale green to grass-green, olive-green
Opacity: Transparent to translucent
Luster: Soapy
Streak: Colorless
SGLow: 2.6
SGHigh: 3.3
HardnessLow: 2
HardnessHigh: 2.5
Cleavage: 1
Direction: {001} perfect
Habit: Tabular six sided crystals, foliated, granular, compact, earthy
Fracture: Earthy, thin flexible flakes
Other: Widespread within metamorphic rocks, especially schists. Alteration pproduct of pyroxenes and olivine
Comments: A group name that resembles green mica

Optical Properties

Name: Chlorite
Group: Chlorite
Formula: $(Mg,Fe)_5Al(Si_3Al)O_{10}(OH)_8$
Crystal System: Monoclinic
Color: Green to pale gren
Form: Scaly masses, cleavage masses, flaky crystals, tabular crystals with pseudo-hexagonal outlines, often bent
Relief: Fair, n>balsam
Birefringence: Weak
2V: 0"-50"
Nalpha or Nord.: 1.570-1.658
NBeta or Nextr.: 1.571-1.667
NGamma: 1.576-1.667
Optical Sign: Biaxial positive
Orientation: Length fast
Pleochroism: Greenish pleochroism
Twinning: Polysynthetic according to mica law
Cleavage: One perfect{001}
Extinction: Parallel upto 9" with cleavage
Alteration:
Features: A group name that resembles green mica. Over 24 sub species have been identified.
Occurrence: Widespread within metamorphic rocks, especially schists. Alteration pproduct of pyroxenes and olivine

Physical Properties

Name: Chloritoid

Group: Mica

Formula: (Fe,Mg,Mn)2Al4Si2O10(OH)4

Crystal System: Monoclinic (and tric.?)

Color: Dark grey, greenish grey to greenish black

Opacity: Translucent

Luster: Pearly on cleavage

Streak: Colorless

SGLow: 3.61

SGHigh: 3.61

HardnessLow: 6.5

HardnessHigh: 6.5

Cleavage: 3

Direction: {001} perfect, {110} distinct, {010} parting

Habit: Crystals tabular, pseudohexagonal, rare; massive, foliated, thin scales or plates

Fracture: Laminae brittle

Other: Occurs within metamorphic schists with mica

Comments: Series with carboirite

Optical Properties

Name: Chloritoid

Group: Mica

Formula: (Fe,Mg,Mn)2Al4Si2O10(OH)4

Crystal System: Monoclinic (and tric.?)

Color: Colorless to green to greenish-grey

Form: Pseudo hexagonal tabular and elongated crystals. Often shows houglass structure due to inclusions

Relief: High, n>balsam

Birefringence: Weak to moderate, 0.013-0.016

2V: 36"-63"

Nalpha or Nord.: 1.715-1.724

NBeta or Nextr.: 1.719-1.726

NGamma: 1.731-1.737

Optical Sign: Biaxial positive

Orientation: Length fast, r>v or r<v

Pleochroism: Usually somewhat pleochroic

Twinning: Polysynthetic, according to mica law

Cleavage: Two, perfect {001}, imperfect {110}

Extinction: Almost parallel upto 20"

Alteration:

Features: Similar to other chlorites, but relief is higher and cleavage less perfect.

Occurrence: Occurs within metamorphic schists with mica

Physical Properties

Name: Chondrodite
Group: Humite
Formula: (Mg,Fe)5(SiO4)2(F,OH)2
Crystal System: Monoclinic
Color: Yellow, brown, red
Opacity: Transparent to translucent
Luster: Vitreous
Streak:
SGLow: 3.16
SGHigh: 3.26
HardnessLow: 6
HardnessHigh: 6.5
Cleavage: 1
Direction: {100} indistinct
Habit: Crystals varied, highly modified; massive
Fracture: Subconchoidal to uneven; brittle
Other: Characteristic of metamorphic limestones
Comments:

Optical Properties

Name: Chondrodite
Group: Humite
Formula: (Mg,Fe)5(SiO4)2(F,OH)2
Crystal System: Monoclinic
Color: Colorless to yellowish to brownish
Form: Subhedral with rounded outlines and in large anhedral crystals
Relief: High, n>balsam
Birefringence: Strong, 0.027-0.035
2V: 70"-90"
Nalpha or Nord.: 1.592-1.643
NBeta or Nextr.: 1.602-1.655
NGamma: 1.621-1.670
Optical Sign: Biaxial positive
Orientation: Length fast,r>v weak
Pleochroism: deep colored varieties are pleochroic from neutral - brown, pale brown-red brown
Twinning: Simple and polysynthetic twinning common {001}
Cleavage: One parting parallel to {001} due to twinning
Extinction: Oblique minus 26" to minus 31" from twin plane
Alteration:
Features: Color, pleochroism distinctive.
Occurrence: Characteristic of metamorphic limestones

Physical Properties

Name: Chromite

Group: Chromite

Formula: $FeCr_2O_4$

Crystal System: Isometric

Color: Iron-black to brownish-black

Opacity: Opaque

Luster: Metallic to sub-metallic

Streak: Dark brown

SGLow: 4.3

SGHigh: 5.0

HardnessLow: 5.5

HardnessHigh: 5.5

Cleavage: None

Direction:

Habit: Octohedral crystals, usually disseminated grains

Fracture: Conchoidal, uneven, brittle

Other: Widespread as an accessory in igneous rocks, particularl in peridotites.

Comments: Weakly magnetic

Optical Properties

Name: Chromite

Group: Oxides

Formula: $(Fe,Mg)(Cr,Al,Fe)_2O_4$

Crystal System: Isometric

Color: Black with submettalic in reflected light

Form: Subhedral grains or aggregates, minute octohedra

Relief: V.high, opaque

Birefringence: Opaque to transluscent on edges

2V:

Nalpha or Nord.:

NBeta or Nextr.:

NGamma:

Optical Sign:

Orientation:

Pleochroism:

Twinning:

Cleavage:

Extinction:

Alteration:

Features: Black color, isotropism

Occurrence: Usually in peridotites, pyroxenites and dunites as a late magmatic mineral

Physical Properties

Name: Chrysotile
Group: Serpentine
Formula: Mg3Si2O5(OH)4
Crystal System: Monoclinic
Color: Yellow, white, grey, green
Opacity: Opaque
Luster: Greasy, waxy, silky
Streak: White
SGLow: 2.55
SGHigh: 2.56
HardnessLow: 2.5
HardnessHigh: 2.5
Cleavage: 1
Direction: None
Habit: Fibrous, lamellar and columnar
Fracture: Splintery, brittle
Other: A metamorphic mineral principally found in serpentinites
Comments: A subgroup name for the sheet silicates of the serpentine group

Optical Properties

Name: Chrysotile
Group: Serpentine
Formula: Mg3Si2O5(OH)4
Crystal System: Monoclinic
Color: Colorless
Form: Cross fibre veinlets (asbestiform)
Relief: Low slightly > balsam
Birefringence: Moderate, 0.011-0.014
2V: 0"-50"
Nalpha or Nord.: 1.493-1.546
NBeta or Nextr.: 1.504-1.550
NGamma: 1.517-1.557
Optical Sign: Biaxial,positive
Orientation: Length slow
Pleochroism:
Twinning:
Cleavage: One perfect {001}
Extinction: Parallel to length
Alteration:
Features: Asbestiform structure distinctive.
Occurrence: A metamorphic mineral principally found in serpentinites

Optical Properties

Name: Cliachite

Group: Bauxite

Formula: Al2O3(H2O)x

Crystal System: Mineraloid

Color: Colorless to deep brown or red

Form: Pisolitic or massive

Relief: Moderate, n>balsam

Birefringence: Nil, Isotropic

2V:

Nalpha or Nord.: 1.57-1.6

NBeta or Nextr.:

NGamma:

Optical Sign:

Orientation:

Pleochroism:

Twinning:

Cleavage: None

Extinction:

Alteration:

Features: Pisolitic structure and association with gibbsite, contraction cracks from shrinkage of original gell form

Occurrence: Main constituent of many bauxites. Forms from

Physical Properties

Name: Clinochlore
Group: Chlorite
Formula: (Mg,Fe)5Al(Si3Al)O10(OH)8
Crystal System: Monoclinic
Color: Colorless, white, yellowish, pale green to grass-green, olive-green, purplish-red
Opacity: Transparent to translucent
Luster: Soapy
Streak: Colorless
SGLow: 2.63
SGHigh: 2.98
HardnessLow: 2
HardnessHigh: 2.5
Cleavage: 1
Direction: {001} perfect
Habit: Crystals tabular, hexagonal cross section; massive, foliated, coarse scaly granular, fine granular, earthy
Fracture: Laminae flexible, inelastic
Other: Widespread within metamorphic rocks, especially schists. Alteration pproduct of pyroxenes and olivine
Comments: Series with chamosite

Optical Properties

Name: Clinochlore
Group: Chlorite
Formula: (Mg,Fe)5Al(Si3Al)O10(OH)8
Crystal System: Monoclinic
Color: Colorless to green
Form: Thin to thick tabular crystals with six sided outlines. Often bent
Relief: Fair, n>balsam
Birefringence: Weak, 0.004-0.011
2V: 0"-50"
Nalpha or Nord.: 1.571-1.588
NBeta or Nextr.: 1.571-1.588
NGamma: 1.576-1.597
Optical Sign: Biaxial positive
Orientation: Length fast, r<v
Pleochroism: Green - colorless pleochrism
Twinning: Polysyntheitc according to mica law
Cleavage: Perfect {001}
Extinction: Oblique, almost parallel 2"-9" along cleavage
Alteration:
Features: Similar to other chlorites, but with oblique extinction. Similar to penninite, but greater birefringence and biaxial figure.
Occurrence: Widespread within metamorphic rocks, especially schists. Characteristic of chlorite schists. Alteration product of pyroxenes and olivine

Physical Properties

Name: Clinozoisite
Group: Epidote
Formula: Ca2Al3(SiO4)3(OH)
Crystal System: Monoclinic
Color: Colorless, pale yellow, light brown, grey, green, pink, pink-red
Opacity: Transparent to translucent
Luster: Vitreous
Streak: Colorless, greyish
SGLow: 3.21
SGHigh: 3.38
HardnessLow: 6.5
HardnessHigh: 6.5
Cleavage: 1
Direction: {001} perfect
Habit: Crystals short to long prismatic, deeply striated; acicular, massive, coarse to fine granular; fibrous, parallel, divergent
Fracture: Uneven; brittle
Other: Widespread mineral in igneous and metamorphic rocks
Comments: Series with epidote; dimorphous with zoisite

Optical Properties

Name: Clinozoisite
Group: Epidote
Formula: Ca2Al3(SiO4)3(OH)
Crystal System: Monoclinic
Color: Colorless
Form: Elongated prisms or columnar aggregates and six sided cross sections
Relief: High, n>balsam
Birefringence: Weak, 0.005-0.011
2V: 66"-90"
Nalpha or Nord.: 1.710-1.723
NBeta or Nextr.: 1.715-1.729
NGamma: 1.719-1.734
Optical Sign: Biaxial positive
Orientation: Length slow or length fast, r<v strong
Pleochroism:
Twinning: Plysynthetic along {100} infrequent
Cleavage: One perfect parallel to {001}
Extinction: Parallel
Alteration:
Features: Similar to zoisite but has distinctive first order yellow green birefringence and larger axial angle. Distinguished from epidote by lack pleochroism, weak birefringence and positive axial sign
Occurrence: Widespread mineral in igneous and metamorphic rocks, similar occurrence as epidote

Physical Properties

Name: Colemantite
Group: Borate
Formula: $Ca_2B_6O_{11}+5H_2O$
Crystal System: Monoclinic
Color: Colorless, white, yellowish white, greyish
Opacity: Transparent to translucent
Luster: Dull
Streak: White
SGLow: 2.42
SGHigh: 2.42
HardnessLow: 4.5
HardnessHigh: 4.5
Cleavage: 2
Direction: {010} perfect, {001} distinct
Habit: Crystals equant, short prismatic, pseudohexagonal; massive; rounded aggregates
Fracture: Subconchoidal to uneven; brittle
Other: Occurs within beds in desert environments
Comments:

Optical Properties

Name: Colemantite
Group: Borate
Formula: $Ca_2B_6O_{11}+5H_2O$
Crystal System: Monoclinic
Color: Colorless, milky white, yellowish, grey or muddy
Form: Short, prismatic crystals, massive and compact
Relief: Moderate to high, n>balsam
Birefringence: Moderate, 0.028
2V: 55"
Nalpha or Nord.: 1.586
NBeta or Nextr.: 1.592
NGamma: 1.614
Optical Sign: Biaxial positive
Orientation: Dispersion slight r>v, optic plane at right angles to {010}
Pleochroism:
Twinning:
Cleavage: Two {010} perfect, {001} fair
Extinction:
Alteration:
Features: Slightly soluble in water, indices of refraction > kernite and borax
Occurrence: Occurs within beds in desert environments

Optical Properties

Name: Collophane

Group: Apatite

Formula: 3Ca3(PO4)2.nCa(CO3)3,F2,O)(H2O)

Crystal System: Amorphous or cryptocrystalline

Color: Light to dark brown, yellowish brown grey, colorless

Form: Massive, oolitic or colloform in grains and fragments of organic material, Organic structure of brachiopods, crinoids etc.

Relief: Moderate, n>balsam

Birefringence: Isotropic, sometimes weak, upto 0.005

2V:

Nalpha or Nord.: 1.57-1.62

NBeta or Nextr.:

NGamma:

Optical Sign:

Orientation: Lenght slow or length fast

Pleochroism:

Twinning:

Cleavage: None, irregular fracture sometimes

Extinction:

Alteration: Alters to calcite. Quartz, chalcedony and opal rare.

Features: Isotropism, form, color

Occurrence: In sedimentary phosphate rocks and limestones. Main mineral in fossils due to phosphate enrichment or replacement of calcareous organisms.

Physical Properties

Name: Cordierite
Group: Tourmaline
Formula: Mg2Al4Si5O18
Crystal System: Orthorhombic
Color: Blue, bluish violet, greenish, grey, yellowish, brown
Opacity: Transparent to translucent
Luster: Vitreous
Streak: Colorless
SGLow: 2.53
SGHigh: 2.78
HardnessLow: 7
HardnessHigh: 7.5
Cleavage: 3
Direction: {010} distinct, {001} and {100} indistinct
Habit: Crystals short prismatic; massive, compact, embedded grains
Fracture: Conchoidal; brittle
Other: Widespread in metamorphic rocks, especially hornfelses, schists and gneisses
Comments: Series with sekaninaite; dimorphous with indialite; may be strongly pleochroic

Optical Properties

Name: Cordierite
Group: Tourmaline
Formula: Mg2Al4Si5O18
Crystal System: Orthorhombic
Color: Colorless
Form: Short prismatic, pseudo-hexagonal crystals. Anhedral grains and aggregates also common. Inclusions with pleochroic haloes common.
Relief: Low, n slightly <balsam or slightly > balsam
Birefringence: Weak, 0.007-0.011
2V: 40"-80"
Nalpha or Nord.: 1.532-1.552
NBeta or Nextr.: 1.536-1.562
NGamma: 1.539-1.570
Optical Sign: Biaxial negative or positive
Orientation: r<v weak
Pleochroism: V. thick sections are pleochroic from yellow - dark violet or blue - pale blue or violet
Twinning: Penetration twins along {110}. Also polysynthetic.
Cleavage: Three {010} distinct, {001} and {100} indistinct. Imperfect {010}, Parting ({001} due to alteration
Extinction: Parallel to outlines
Alteration: Alters to sericite
Features: Very similar to quartz, but is biaxial and often shows twinning.
Occurrence: Widespread in metamorphic rocks, especially hornfelses, schists and gneisses

Physical Properties

Name: Corundum

Group: Corundum

Formula: Al2O3

Crystal System: Trigonal

Color: Colorless, grey, brown, blue, red, green, yellow, orange, purple

Opacity: Transparent to translucent

Luster: Vitreous to adamantine

Streak: Colorless

SGLow: 4.0

SGHigh: 4.1

HardnessLow: 9

HardnessHigh: 9

Cleavage: 2

Direction: {0001} and {10-11} parting

Habit: Crystals often well-developed, steep-pyramidal, prismatic, tabular; barrel-shaped; massive

Fracture: Conchoidal

Other: Cryptocrystalline variety of quartz

Comments: Sometimes fluoresces under SW and LW

Optical Properties

Name: Corundum

Group: Corundum

Formula: Al2O3

Crystal System: Trigonal

Color: Colorless

Form: Euhedral tabular to prismatic cystals common, cross sections are six sided and may show zoning.

Relief: Very high, n>balsam

Birefringence: Weak, 0.008-0.009

2V: 4.1

Nalpha or Nord.: 1.759 to 1.763

NBeta or Nextr.: 1.767 to 1.772

NGamma:

Optical Sign: Uniaxial neg

Orientation: Length slow (tabular), length fast (prismatic)

Pleochroism: May be pleochroic in thick sections

Twinning: Twinning lamellae wit {101_1} as twin plane

Cleavage: Two {0001} and {10-11} parting. Parting oftem parallel to rhombohedron {10_11} or pinacoid {0001} or both

Extinction:

Alteration: Unusual

Features: Very high relief with weak birefringence, parting and twin lamellae

Occurrence: Cryptocrystalline variety of quartz, Characteristic of corundum syenites, contact metamorphic limestones and metamorphic shales

Physical Properties

Name: Cristabolite
Group: Quartz
Formula: SiO2
Crystal System: Tetragonal
Color: White, grey, bluish grey, yellowish, brownish
Opacity: Translucent to opaque
Luster: Vitreous
Streak: Colorless
SGLow: 2.33
SGHigh: 2.33
HardnessLow: 6.5
HardnessHigh: 6.5
Cleavage: 0
Direction: None
Habit: Crystals pseudo-octahredral or rarely pseudocubic, small; massive; stalactitic, spherulitic, crusts or botryoidal
Fracture: Brittle, conchoidal
Other: Usually occurs within cavities of volcanic igneous rocks
Comments: Polymorphous with coesite, quartz, stishovite and tridymite

Optical Properties

Name: Cristabolite (Crystabolite)
Group: Quartz
Formula: SiO2
Crystal System: Tetragonal
Color: Colorless
Form: Minute square crystals in cavities of volcanic rocks. Intergrowths with feldspar fibres in spherulites
Relief: Moderate, n<balsam
Birefringence: Weak, 0.003
2V:
Nalpha or Nord.: 1.484
NBeta or Nextr.: 1.487
NGamma:
Optical Sign:
Orientation:
Pleochroism:
Twinning:
Cleavage: Curved fracture that is distinctive
Extinction:
Alteration: Polymorphous with quartz and tridymite
Features: Curved fracture distinctive, may require RI determination.
Occurrence: Usually occurs within cavities of volcanic igneous rocks

Physical Properties

Name: Crossite

Group: Amphibole

Formula: Na2(Mg,Fe)3(Al,Fe)2Si8O22(OH)2

Crystal System: Monoclinic

Color: Blue, greyish

Opacity: Translucent

Luster: Vitreous to dull

Streak:

SGLow: 3.11

SGHigh: 3.211

HardnessLow: 6

HardnessHigh: 6

Cleavage: 1

Direction: {110} perfect

Habit: Crystals prismatic, often lath-like; massive, fibrous, columnar, granular

Fracture: Uneven; brittle

Other:

Comments:

Physical Properties

Name: Cummingtonite

Group: Amphibole

Formula: (Mg,Fe)7Si8O22(OH)2

Crystal System: Monoclinic

Color: Dark green, greyish green, brown; white, light grey

Opacity: Translucent to nearly opaque

Luster: Silky

Streak:

SGLow: 3.10

SGHigh: 3.47

HardnessLow: 5

HardnessHigh: 6

Cleavage: 1

Direction: {110} good

Habit: Fibrous or fibro-lamellar, radiated

Fracture:

Other: Characteristic of metamorphic schists and hornfels

Comments: Series with magnesiocummingtonite and grunerite

Optical Properties

Name: Cummingtonite

Group: Amphibole

Formula: (Mg,Fe)7Si8O22(OH)2

Crystal System: Monoclinic

Color: Colorless to nuetral

Form: Parallel to slightly radiating aggregates of prismatic crystals

Relief: Moderate, n>balsam

Birefringence: Strong, 0.025-0.029

2V: 68"-87"

Nalpha or Nord.: 1.639-1.657

NBeta or Nextr.: 1.645-1.669

NGamma: 1.664-1.686

Optical Sign: Biaxial positive

Orientation: Length slow, r<v

Pleochroism: May show weak pleochroism

Twinning: Polysynthetic twins along {100} are characteristic

Cleavage: In two directions 56" and 124", as in other amphiboles

Extinction: Oblique 15"-20" in longitudinal sections.

Alteration:

Features: Series with magnesiocummingtonite and grunerite. Larger extinction angle than grunerite. Higher RI than tremolite.

Occurrence: Characteristic of metamorphic schists and hornfels

Physical Properties

Name: Diaspore
Group: Bauxite
Formula: AlO(OH)
Crystal System: Orthorhombic
Color: White to colorless, yellowish, greenish, lilac, pink, brownish
Opacity: Transparent to subtranslucent
Luster: Vitreous, pearly on cleavages
Streak: White
SGLow: 3.3
SGHigh: 3.5
HardnessLow: 6.5
HardnessHigh: 7
Cleavage: 3
Direction: {010} perfect, {110} distinct, {100} traces
Habit: Crystals thin elongated plates; acicular or tabular, striated; massive, foliated, scaly, stalactitic
Fracture: Conchoidal; brittle
Other: Occurs in metamorphic schists, altered igneous rocks and flint clays
Comments: Dimorphous with boehmite

Optical Properties

Name: Diaspore
Group: Bauxite
Formula: AlO(OH)
Crystal System: Orthorhombic
Color: Colorless to pale blue
Form: Tabular parallel to {010}. Also occurs in mineral aggregates
Relief: High, n>balsam
Birefringence: Strong 0.048
2V: 84"
Nalpha or Nord.: 1.702
NBeta or Nextr.: 1.722
NGamma: 1.750
Optical Sign: Biaxial positive
Orientation: Length fast, dispersion, r<v weak
Pleochroism: May be pleochroic in thick sections
Twinning:
Cleavage: Perfect, one direction {010}
Extinction: Parallel
Alteration: Dimorphous with boehmite
Features: Cleavage, crystals and relief
Occurrence: Occurs in metamorphic schists, altered igneous rocks and flint clays

Physical Properties

Name: Dickite

Group: Kaolinite, Clay

Formula: Al2Si2O5(OH)4

Crystal System: Monoclinic

Color: Colorless, white

Opacity: Transparent to translucent

Luster: Satiny to dull earthy

Streak: White

SGLow: 2.60

SGHigh: 2.60

HardnessLow: 2

HardnessHigh: 2.5

Cleavage: 1

Direction: {001} perfect

Habit: Crystals thin tabular, pseudohexagonal, minute; stacked platelets, massive, compact, friable or mealy

Fracture: Scales flexible, inelastic

Other: Usually found as a diagenetic clay mineral in sedimentary beds

Comments: Polymorphous with halloysite, kaolinite and nacrite

Optical Properties

Name: Dickite

Group: Kaolinite, Clay

Formula: Al2Si2O5(OH)4

Crystal System: Monoclinic

Color: Colorless to pale yellow

Form: Microcrystalline pseudo-hexagonal plates

Relief: Low, n>balsam

Birefringence: Weak, 0.006

2V: 52"-80"

Nalpha or Nord.: 1.560

NBeta or Nextr.: 1.562

NGamma: 1.566

Optical Sign: Biaxial positive

Orientation: Length slow

Pleochroism:

Twinning:

Cleavage: One perfect {001}

Extinction: Oblique 15"-20" on base

Alteration:

Features: Similar to kaolinite but larger extinction angle and slightly higher birefringence.

Occurrence: Usually found as a diagenetic clay mineral in sedimentary beds.

Physical Properties

Name: Diopside

Group: Pyroxene

Formula: $CaMgSi_2O_6$

Crystal System: Monoclinic

Color: Colorless, white, grey, pale green to dark greenish black, brown, blue, rose

Opacity: Transparent to nearly opaque

Luster: Vitreous, often dull

Streak: White or greyish

SGLow: 3.22

SGHigh: 3.38

HardnessLow: 5.5

HardnessHigh: 6.5

Cleavage: 3

Direction: {110} good, {100} and {010} parting

Habit: Crystals short prismatic; massive, columnar, granular, lamellar

Fracture: Uneven

Other: Characteristic of contact metamorphic zones

Comments: Series with hedenbergite and with johannsenite, from many rocks and meteorites

Optical Propeties

Name: Diopside

Group: Pyroxene

Formula: $CaMgSi_2O_6$

Crystal System: Monoclinic

Color: Colorless, neutral, pale green to bright green

Form: Subhedral prismatic crystals. Cross sections four or eight sided

Relief: High, n>balsam

Birefringence: 3.22

2V: 58"-60"

Nalpha or Nord.: 1.650-1.698

NBeta or Nextr.: 1.657-1.706

NGamma: 1.681-1.727

Optical Sign: Biaxial positive

Orientation: Length slow, r<v weak

Pleochroism:

Twinning: Twinning along {100} and and secondary polysynthetic twinning along {001} common

Cleavage: Three distinct {110} and in two directions 87" and 93" in cross sections

Extinction: Oblique minus 37" - minus 44" parallel to c-axis and symmetrical in cross sections

Alteration: lters to tremolite/actinolite

Features: Pale green color and extinction angle

Occurrence: Characteristic of contact metamorphic zones

Physical Properties

Name: Dolomite
Group: Dolomite
Formula: CaMg(CO3)2
Crystal System: Trigonal
Color: Colorless, white, greyish, greenish, pale brown, pinkish
Opacity: Transparent to subtranslucent
Luster: Vitreous to pearly
Streak: White
SGLow: 2.85
SGHigh: 2.85
HardnessLow: 3.5
HardnessHigh: 4
Cleavage: 1
Direction: {10-11} perfect
Habit: Crystals simple rhombohedrons, often with curved faces, rarely tabular or octahedral; crystalline aggregates; massive
Fracture: Subconchoidal; brittle
Other: The major constituent of sedimentary and metamorphic dolomite rock. Also occurs in veins and hydrothermal deposits
Comments: Series with ankerite and with kutnohorite

Optical Properties

Name: Dolomite
Group: Dolomite
Formula: CaMg(CO3)2
Crystal System: Hexagonal
Color: Colorless to grey
Form: Fine to coarse subhedral grains. Euhedral, rhombohedral grains. Zonal structure common duet variation in iron content.
Relief: High in long direction of rhomb, low in short direction.
Birefringence: Extreme, 0.180-0.190
2V:
Nalpha or Nord.: 1.500-1.526
NBeta or Nextr.: 1.680-1.716
NGamma:
Optical Sign: Uniaxial negative
Orientation:
Pleochroism:
Twinning: Polysynthetic twinning along {022_1}
Cleavage: One perfect rhombohedral parallel to {101_1}
Extinction: Symmetrical to cleavage traces, curved crsytals show wavy extinction
Alteration:
Features: Similar to calcite and magnesite. Distinguished from calcite by euhedral crystal form, zonal structure and twinning lamellae parallel to short diagonal. Staining may be required to differentiate magnesite.
Occurrence: The major constituent of sedimentary and metamorphic dolomite rock. Also occurs in veins and hydrothermal deposits

Physical Properties

Name: Dravite (Tourmailine Group)

Group: Tourmaline

Formula: NaMg3Al6(BO3)3Si6O18(OH)4

Crystal System: Trigonal

Color: Brown, black, greenish black, dark red, pale bluish green to emerald-green

Opacity: Transparent to nearly opaque

Luster: Vitreous to somewhat resinous

Streak: Colorless

SGLow: 3.03

SGHigh: 3.15

HardnessLow: 7

HardnessHigh: 7

Cleavage: 2

Direction: {11-20} and {10-11} indistinct

Habit: Crystals short to long prismatic; sometimes nearly equant; single crystals or radiating groups; massive, compact, grains

Fracture: Conchoidal to uneven; brittle

Other: Mainly found within ggranite pegmatites, granites and high temperature veins

Comments: Series with schorl and with elbaite

Optical Properties

Name: Dravite (Tourmailine Group)

Group: Tourmaline

Formula: NaMg3Al6(BO3)3Si6O18(OH)4

Crystal System: Trigonal

Color: Colorless to pale yellow

Form: Prismatic with alongated, hexagonal or triangular outlines in cross section, columnar and fibrous radiating aggregates

Relief: High, n>balsam

Birefringence: Moderate, 0.022-0.040

2V:

Nalpha or Nord.: 1.613-1.628

NBeta or Nextr.: 1.632-1.655

NGamma:

Optical Sign: Uniaxial negative

Orientation: Length fast

Pleochroism: Weak pleochroism

Twinning:

Cleavage: None, irregular fractures frequent

Extinction: Parallel

Alteration:

Features: Distinguished from other tourmailines by pale yellow color. Elongated and hexagonal outlines, color, pleochroism, parallel extinction and lack of cleavage distinctive.

Occurrence: Dravite mainly found within metamorphic limestones and some schists.

Physical Properties

Name: Dumortierite
Group: Silliminite
Formula: Al7O3(BO)3(SiO4)3
Crystal System: Orthorhombic
Color: Violet, blue, rarely pink, brown
Opacity: Translucent
Luster: Vitreous, dull
Streak: White
SGLow: 3.3
SGHigh: 3.4
HardnessLow: 7
HardnessHigh: 7
Cleavage: Good, one direction, poor second direction
Direction:
Habit: Usually in fibrous and columnar masses, crystals rare
Fracture: Uneven, hackly
Other: In granite pegmatites with quartz and andalusite and in hydrothermal replacement deposits
Comments:

Optical Properties

Name: Dumortierite
Group: Sillimanite
Formula: HBAl8Si3O20
Crystal System: Orthorhombic
Color: Colorless to blue, lavendar, pink and reddish
Form: Long prismatic to acicular crystals and felted masses
Relief: High, n>balsam
Birefringence: Weak to moderate, 0.011-0.020
2V: 20"-40"
Nalpha or Nord.: 1.659-1.678
NBeta or Nextr.: 1.684-1.691
NGamma: 1.686-1.692
Optical Sign: Biaxial negative
Orientation: Length fast, r>v
Pleochroism:
Twinning: Occasional penetration twins with {110}
Cleavage: Two imperfect parallel to length. Also cross fractures
Extinction: Parallel
Alteration: Alters to sericite
Features: Color and pleochroism distinctive
Occurrence: Occurs in metamorphic rocks, including, schists, gneisses and grnite pegmatites.

Physical Properties

Name: Elbaite (Tourmaline Group)

Group: Tourmaline

Formula: Na(Li,Al)3Al6(BO3)3Si6O18(OH)4

Crystal System: Trigonal

Color: Green, blue, red, yellow, white, colorless

Opacity: Transparent to translucent

Luster: Vitreous

Streak: Colorless

SGLow: 3.03

SGHigh: 3.10

HardnessLow: 7

HardnessHigh: 7

Cleavage: 2

Direction: {11-20} and {10-11} indistinct

Habit: Crystals short to long prismatic, vertically striated, hemimorphic; massive, compact, columnar to fibrous

Fracture: Conchoidal to uneven; brittle

Other: Mainly found within ggranite pegmatites, granites and high temperature veins

Comments: Series with dravite, typically color zoned

Optical Properties

Name: Elbaite (Tourmaline Group)

Group: Tourmaline

Formula: Na(Li,Al)3Al6(BO3)3Si6O18(OH)4

Crystal System: Trigonal

Color: Colorless

Form: Prismatic with alongated, hexagonal or triangular outlines in cross section, columnar and fibrous radiating aggregates. Elbaite may be spherulitic

Relief: High, n>balsam

Birefringence: Moderate, 0.19-0.025

2V:

Nalpha or Nord.: 1.615-1.629

NBeta or Nextr.: 1.635-1.655

NGamma:

Optical Sign: Uniaxial negative

Orientation: Length fast

Pleochroism: Thick sectionms may be pleochroic from coloorless - pink or pale green or pale blue

Twinning:

Cleavage: None, irregular fractures frequent

Extinction: Parallel, dark in cross sections

Alteration:

Features: Similar to other tourmalines, pink and blue colors and occurrence may distinguish

Occurrence: Elbaite mainly found within granite pegmatites, granites and high temperature veins

Physical Properties

Name: Enstatite

Group: Pyroxene

Formula: Mg2Si2O6

Crystal System: Orthorhombic

Color: Colorless, yellowish or greenish white, grey, olive-green, brown

Opacity: Transparent to nearly opaque

Luster: Vitreous to pearly

Streak: Colorless, greyish

SGLow: 3.21

SGHigh: 3.43

HardnessLow: 5

HardnessHigh: 6

Cleavage: 3

Direction: {210} good, {100} and {010} parting

Habit: Crystals prismatic; massive, lamellar, fibrous

Fracture: Uneven; brittle

Other: Characteristic of mafic igneous rocks and serpentinites

Comments: Series with orthoferrosilite and with hypersthene; dimorphous with clinoenstatite

Optical Properties

Name: Enstatite

Group: Pyroxene

Formula: Mg2Si2O6

Crystal System: Orthorhombic

Color: Colorless to nuetral

Form: Prismatic with pyroxene cross section. Inclusions exhibit schiller structure in bronzite variety

Relief: High, n>balsam

Birefringence: Weak, 0.008-0.009, pale yellow of first order strongest

2V: 58"-80"

Nalpha or Nord.: 1.650-1.665

NBeta or Nextr.: 1.653-1.670

NGamma: 1.658-1.674

Optical Sign: Biaxial positive

Orientation: Length slow, r<v weak

Pleochroism:

Twinning: Rare

Cleavage: Perfect in two directions {110} and nearly at right angles, 88" and 92". In longitudinal sections, cleavage in one direction parallel to outlines. Parting parallel to {010} on occasion

Extinction: Parallel

Alteration: Alters to antigorite

Features: Parallel extinction and lack of pleochroism

Occurrence: Characteristic of mafic igneous rocks and serpentinites. Also found in meteorites

Physical Properties

Name: Epidote

Group: Epidote

Formula: Ca2(Al,Fe)3(SiO4)3(OH)

Crystal System: Monoclinic

Color: Yellowish green to brownish to dark green, grey, greyish white, greenish black, black

Opacity: Transparent to nearly opaque

Luster: Vitreous to pearly

Streak: Colorless to greyish

SGLow: 3.35

SGHigh: 3.5

HardnessLow: 6

HardnessHigh: 7

Cleavage: 1

Direction: {001} perfect

Habit: Crystals short to long prismatic, may be deeply striated; thick tabular; acicular; massive, coarse to fine granular; fibrous

Fracture: Uneven; brittle

Other: Widespread mineral in igneous and metamorphic rocks

Comments: Series with clinozoisite

Optical Properties

Name: Epidote

Group: Epidote

Formula: Ca2(Al,Fe)3(SiO4)3(OH)

Crystal System: Monoclinic

Color: Colorless to yellow-green

Form: Elongated prismatic tabular and bladed crystals and granular to columnar aggregates. Six sided cross sections.

Relief: high, n>balsam

Birefringence: Moderate to strong, 0.014-0.045

2V: 69"-89"

Nalpha or Nord.: 1.720-1.734

NBeta or Nextr.: 1.724-1.763

NGamma: 1.734-1.779

Optical Sign: Biaxial negative

Orientation: Length slow or length fast, r>v

Pleochroism: Weak-moderate pleochroism

Twinning: Infrequent polysynthetic along {100}

Cleavage: One perfect parallel to {001}

Extinction: Parallel in longitudinal sectoions

Alteration:

Features: Similar to zoisite and clinozoisite but with strong birefringence.

Occurrence: Widespread mineral in igneous and metamorphic rocks. Deuteric or late magmatic mineral in igneous rocks.

Physical Properties

Name: Erionite (Zeolite Group)
Group: Zeolite
Formula: $(K_2,Ca,Na_2)2Al_4Si_{14}O_{36}+15H_2O$
Crystal System: Hexagonal
Color: White
Opacity: Transparent to translucent
Luster: Vitreous and pearly
Streak: Colorless
SGLow: 2.02
SGHigh: 2.02
HardnessLow: 3.5
HardnessHigh: 4
Cleavage:
Direction:
Habit: Crystals prismatic, minute, in radiating groups; finely fibrous and wool-like
Fracture: Uneven, brittle
Other: In cavities and veins within lavas, particularly basalts
Comments:

Optical Properties

Name: Erionite (Zeolite Group)
Group: Zeolite
Formula: $(K_2,Ca,Na_2)2Al_4Si_{14}O_{36}+15H_2O$
Crystal System: Hexagonal
Color: Colorless
Form: Fibrous, also prismatic aggregates
Relief: Low, n=balsam
Birefringence: Moderate, 0.015
2V:
Nalpha or Nord.: 1.530
NBeta or Nextr.: 1.533
NGamma:
Optical Sign:
Orientation:
Pleochroism:
Twinning:
Cleavage:
Extinction: Probably parallel
Alteration:
Features: Similar to other zeolites
Occurrence: In cavities and veins within lavas, particularly basalts

Physical Properties

Name: Fayalite (Olivine)

Group: Olivine

Formula: Fe2SiO4

Crystal System: Orthorhombic

Color: Greenish yellow, yellowish brown, brown

Opacity: Transparent to translucent

Luster: Vitreous to greasy

Streak: Colorless

SGLow: 4.32

SGHigh: 4.32

HardnessLow: 7

HardnessHigh: 7

Cleavage: 2

Direction: {010} and {100} imperfect

Habit: Crystals thick tabular, often with wedge-shaped terminations, small; massive, compact or granular

Fracture: Conchoidal; brittle

Other:

Comments: Series with forsterite and with tephroite

Optical Properties

Name: Fayalite (Iron Olivine)

Group: Olivine

Formula: Fe2SiO4

Crystal System: Orthorhombic

Color: Colorless to yellow to neutral

Form: Usually anhedral, but may be euhedral in cavities. Polygonal outlines and tabular crystals

Relief: V. high, n>balsam

Birefringence: Strong, 0.042-0.051

2V: 47"-54"

Nalpha or Nord.: 1.805-1.835

NBeta or Nextr.: 1.838-1.877

NGamma: 1.847-1.886

Optical Sign: Biaxial negative

Orientation: Length slow, r>v

Pleochroism: May show faint pleochroism

Twinning:

Cleavage: Two {010} and {100} imperfect

Extinction: Parallel to cleavage

Alteration: Alters to grunerite

Features: Iron rich olivine. Biaxial negative, high RI and occurrence distinctive.

Occurrence: Quite rare, widely distributed in lithophysae of rhyolitic obsidian.

Physical Properties

Name: Ferrohornblende (Barkevikite)
Group: Amphibole
Formula: Ca2(Fe,Mg)4Al(Si7Al)O22(OH,F)2
Crystal System: Monoclinic
Color: Green to dark green, greenish brown, black
Opacity: Transparent to opaque
Luster: Vitreous
Streak: Colorless
SGLow: 3.41
SGHigh: 3.41
HardnessLow: 5
HardnessHigh: 6
Cleavage: 3
Direction: {110} perfect, {001} and {100} parting
Habit: Crystals long to short prismatic; massive, compact, granular, columnar, bladed or fibrous
Fracture: Subconchoidal to uneven; brittle
Other: Occurs mainly in alkali igneous rocks
Comments: Series with magnesiohornblende

Optical Properties

Name: Ferrohornblende (Barkevikite)
Group: Amphibole
Formula: Ca2(Fe,Mg)4Al(Si7Al)O22(OH,F)2
Crystal System: Monoclinic
Color: Colorless
Form: Crystals long to short prismatic; massive, compact, granular, columnar, bladed or fibrous
Relief: High, n>balsam
Birefringence:
2V: 40"-50"
Nalpha or Nord.: 1.614-1.675
NBeta or Nextr.: 1.618-1.691
NGamma: 1.633-1.701
Optical Sign: Biaxial negative
Orientation: r>v distinct
Pleochroism: Strong yellow brown-red brown-dark brown
Twinning:
Cleavage: Two at 56" and 124"
Extinction: Oblique 11"-18" in longitudinal sections, symmetrical in cross sections
Alteration: Alters to chlorite and epidote
Features: Amphibole cross section, color and pleochroism distinctive
Occurrence: Occurs mainly in alkali igneous rocks with nepheline

Physical Properties

Name: Fluorapatite (Apatite)

Group: Apatite

Formula: Ca5(PO4)3F

Crystal System: Hexagonal

Color: Colorless, white, grey, yellow to yellowish green, green, blue, violet, red, brown

Opacity: Transparent to opaque

Luster: Vitreous to subresinous, silky

Streak: White

SGLow: 3.1

SGHigh: 3.2

HardnessLow: 5

HardnessHigh: 5

Cleavage: 2

Direction: {0001} indistinct, {10-10} trace

Habit: Crystals short to long prismatic, thin to thick tabular or complex; massive compact to coarse granular; fibrous

Fracture: Conchoidal to uneven; brittle

Other:

Comments: Often fluoresces, phosphorescent in UV light, or thermoluminescent

Optical Properties

Name: Fluorapatite (Apatite)

Group: Sillimanite

Formula: Ca5(PO4)3F

Crystal System: Hexagonal

Color: Colorless

Form: Crystals short to long prismatic, thin to thick tabular or complex; massive compact to coarse granular; fibrous

Relief: Moderate, n>balsam

Birefringence: Weak, 0.003-0.004

2V:

Nalpha or Nord.: 1.630-1.651

NBeta or Nextr.: 1.633-1.655

NGamma:

Optical Sign: Uniaxial negative

Orientation: Length fast, tabular sections are length slow

Pleochroism:

Twinning:

Cleavage: Two, imperfect basal {0001} may show as cross fractures, {101_0} may show parallel to length

Extinction: Parallel

Alteration:

Features: Often fluoresces, phosphorescent in UV light, or thermoluminescent. Hexagonal form distinctive

Occurrence:

Physical Properties

Name: Fluorite
Group: Fluorite
Formula: CaF2
Crystal System: Isometric, cubic
Color: Colorless, purple, blue, green, yellow, white, pink, red, brown, bluish black
Opacity: Transparent to translucent
Luster: Vitreous
Streak: White
SGLow: 3.18
SGHigh: 3.18
HardnessLow: 4
HardnessHigh: 4
Cleavage: 1
Direction: {111} perfect
Habit: Crystals cubes or octahedrons, rarely dodecahedrons; massive, coarse to fine granular; botryoidal, fibrous
Fracture: Subconchoidal to splintery; brittle
Other: Mostly in veins associated with granites
Comments: Often fluoresces blue, yellow, white, reddish or pale violet, may thermoluminesce

Optical Properties

Name: Fluorite
Group: Fluorite
Formula: CaF2
Crystal System: Isometric, cubic
Color: Colorless, purple bands or sppots
Form: Sometimes euhedral with cubic outlines, usually anhedral
Relief: Fairly high, n<balsam
Birefringence: Nil, dark between crossed polars
2V:
Nalpha or Nord.: 1.434
NBeta or Nextr.:
NGamma:
Optical Sign:
Orientation:
Pleochroism:
Twinning:
Cleavage: One {111} perfect. Perfect octohedral cleavage. Usually visble as two intersecting lines at angles of 70" and 110", or three intersecting ;lines at 60" and 120"
Extinction:
Alteration:
Features: High relief, perfect octohedral cleavage, isometric, purple spots or bands.
Occurrence: Mostly in veins associated with granites

Physical Properties

Name: Forsterite (Olivine)

Group: Olivine

Formula: Mg2SiO4

Crystal System: Orthorhombic

Color: Green, lemon-yellow, white

Opacity: Transparent to translucent

Luster: Vitreous to greasy

Streak: Colorless

SGLow: 3.28

SGHigh: 3.28

HardnessLow: 7

HardnessHigh: 7

Cleavage: 2

Direction: {010} and {100} imperfect

Habit: Crystals thick tabular, vertically striated; massive, compact or granular; embedded grains

Fracture: Conchoidal; brittle

Other: Characteristic of contact metamorphic zones and metamorphic limestones

Comments: Series with fayalite; trimorphous with ringwoodite and wadsleyite

Optical Properties

Name: Forsterite (Olivine)

Group: Olivine

Formula: Mg2SiO4

Crystal System: Orthorhombic

Color: Colorless

Form: Subhedral to euhedral tabular crystals with rectangular outlines

Relief: High, n>balsam

Birefringence: Strong, 0.035-0.040

2V: 85"-90"

Nalpha or Nord.: 1.635-1.640

NBeta or Nextr.: 1.651-1.660

NGamma: 1.670-1.680

Optical Sign: Biaxial positive

Orientation: Length slow, r<v

Pleochroism:

Twinning:

Cleavage: Two {010} and {100} imperfect, irregular fractures

Extinction: Parallel to outlines and cleavage

Alteration: Alters to antigorite (serpentine)

Features: Similar to regular olivine. Tabular crystals and lower RI. Occurrence distinctive.

Occurrence: Characteristic of contact metamorphic zones and metamorphic limestones

Physical Properties

Name: Fuschite
Group: Mica
Formula: KAl2(Si3Al)O10(OH,F)2
Crystal System: Monoclinic ps. hex.
Color: Green
Opacity: Transparent to translucent
Luster: Vitreous to pearly or silky
Streak: Colorless
SGLow: 2.77
SGHigh: 2.88
HardnessLow: 2.5
HardnessHigh: 4
Cleavage: 1
Direction: {001} perfect
Habit: Crystals tabular, hexagonal or diamond-shaped cross section; massive, scaly or lamellar; plumose or stellate aggregates
Fracture: Thin laminae flexible and elastic
Other:
Comments: Variety of muscovite, not a recognized name

Optical Properties

Name: Fuschite
Group: Mica
Formula: KAl2(Si3Al)O10(OH,F)2
Crystal System: Monoclinic ps. hex.
Color: Colorless
Form: Crystals tabular, hexagonal or diamond-shaped cross section; massive, scaly or lamellar; plumose or stellate aggregates
Relief:
Birefringence:
2V:
Nalpha or Nord.:
NBeta or Nextr.:
NGamma:
Optical Sign:
Orientation:
Pleochroism:
Twinning:
Cleavage:
Extinction:
Alteration:
Features: Variety of muscovite, not a recognized name
Occurrence:

Physical Properties

Name: Gibbsite (Bauxite)

Group: Bauxite

Formula: Al(OH)3

Crystal System: Monoclinic

Color: White, greyish, greenish, reddish white

Opacity: Transparent to translucent

Luster: Vitreous, pearly on cleavage

Streak: White

SGLow: 2.40

SGHigh: 2.40

HardnessLow: 2.5

HardnessHigh: 3.5

Cleavage: 1

Direction: {001} perfect

Habit: Crystals tabular, hexagonal to large size; massive, chalcedony-like coatings and crusts; stalactitic, concretionary, fibrous

Fracture: Tough

Other: Occurs as a surface weathering product from tropical weathering

Comments: Polymorphous with bayerite, doyleite, and nordstrandite

Optical Properties

Name: Gibbsite (Bauxite)

Group: Bauxite

Formula: Al(OH)3

Crystal System: Monoclinic

Color: Colorless to pale brown

Form: Minute pseudo-hexagonal crystals in cavities and in fine crystalline aggregates that are pseudomorphous after feldspar

Relief: Moderate, n>balsam

Birefringence: Moderate, 0.022

2V: 0-40"

Nalpha or Nord.: 1.554-1.567

NBeta or Nextr.: 1.554-1.567

NGamma: 1.576-1.589

Optical Sign: Biaxial positive

Orientation: Length slow

Pleochroism:

Twinning: Polysynthetic twinning with {001}

Cleavage: One direction parallel to {001}

Extinction: Oblique, upto 26"

Alteration: Polymorphous with bayerite, doyleite and nordstrandite

Features: Aggregate structure, birefringence

Occurrence: Occurs as a surface weathering product from tropical weathering in bauxite deposits

Optical Properties

Name: Glass (Volcanic)

Group: Mineraloid

Formula: SiO2

Crystal System: Amorphous

Color: Colorless

Form: Amorphous silica glass, may be banded or show flow structure, vesicular, perlitic, often contains spherulites, microlites, crystallites, microphenocrysts and phenocrysts

Relief: Low to moderate, n<balsam

Birefringence: None, may show weak birefringence

2V:

Nalpha or Nord.: 1.458-1.462

NBeta or Nextr.:

NGamma:

Optical Sign:

Orientation:

Pleochroism:

Twinning:

Cleavage: None

Extinction:

Alteration: Devitified, feldspars, trydimite, cristabolite or monmorilonite from devitrification

Features: Isotropism, low relief and birefringence. Amorphous silica glass, may be banded or show flow structure, vesicular, perlitic, often contains spherulites, microlites, crystallites, microphenocrysts and phenocrysts.

Occurrence: Occurs frequently in the groundmass of volcanic igneous rocks. Often occurs as a rock type, such as obsidian, pumice, perlite or pitchstone. Usually rhyolitic in composition.

Optical Properties

Name: Glauconite

Group: Serpentine

Formula: KMg(Fe,Al)(SiO3)6.3H2O

Crystal System: Monoclinic

Color: Green, yellow-green, olive green

Form: Fine grains and pellets

Relief: Moderate, n>balsam

Birefringence: Moderate to strong, 0.020-0.032

2V: 16"-30"

Nalpha or Nord.: 1.590-1.612

NBeta or Nextr.: 1.609-1.643

NGamma: 1.610-1.644

Optical Sign: Biaxial negative

Orientation: Length slow, r>v

Pleochroism: Pleochroic from yellow-green

Twinning:

Cleavage: One perfect {001}

Extinction: Parallel to cleavage, upto 3"

Alteration: Alters to limonite

Features: Similar to chamosite but color, pleochroism, pellets and grains distinctive. Chamosite shows oolitic structure.

Occurrence: A diagenetic mineral found in marine sediments and characteristic of green sands. Usually indicative of marine conditions

Physical Properties

Name: Glaucophane
Group: Amphibole
Formula: Na2(Mg,Fe)3Al2Si8O22(OH)2
Crystal System: Monoclinic
Color: Greyish, bluish black, lavender, blue, azure blue
Opacity: Translucent
Luster: Vitreous to dull, pearly
Streak: Greyish blue
SGLow: 3.08
SGHigh: 3.15
HardnessLow: 6
HardnessHigh: 6
Cleavage: 1
Direction: {110} perfect
Habit: Crystals slender prismatic; massive; fibrous, columnar, granular
Fracture: Conchoidal to uneven; brittle
Other: High grade metamorphic mineral mainly within glaucophane schists
Comments: Series with ferroglaucophane

Optical Properties

Name: Glaucophane
Group: Amphibole
Formula: Na2(Mg,Fe)3Al2Si8O22(OH)2
Crystal System: Monoclinic
Color: Blue to violet
Form: Prismatic crystals and columnar aggregates. Pseudohexagonal and rhombic cross sections
Relief: High, n>balsam
Birefringence: Moderate, 0.013-0.018
2V: 0"-68"
Nalpha or Nord.: 1.621-1.655
NBeta or Nextr.: 1.638-1.664
NGamma: 1.639-1.668
Optical Sign: Biaxial negative
Orientation: Length slow
Pleochroism:
Twinning:
Cleavage: Two directions {110} at 56" and 124", Symmetrical in cross sections
Extinction: Oblique 4"-6" in longitudinal sections, Symmetrical in cross sections
Alteration:
Features: Color, pleachroism and amphibole cross sections are distinctive.
Occurrence: High grade metamorphic mineral mainly within glaucophane schists

Physical Properties

Name: Goethite
Group: Limonite-Goethite
Formula: FeO(OH)
Crystal System: Orthorhombic
Color: Blackish brown, reddish or yellowish brown, brownish yellow
Opacity: Opaque
Luster: Adamantine-metallic to dull
Streak: Orange to brownish yellow
SGLow: 3.3
SGHigh: 4.3
HardnessLow: 5
HardnessHigh: 5.5
Cleavage: 2
Direction: {010} perfect, {100} distinct
Habit: Crystals prismatic, vertically striated, capillary or acicular in radiating clusters; massive radial, ocherous, colloform, compact
Fracture: Uneven; brittle
Other: Secondary iron oxide, usually the product of weathering
Comments: Polymorphous with akaganeite, feroxyhyte, and lepidocrocite

Optical Properties

Name: Goethite
Group: Limonite-Goethite
Formula: FeO(OH)
Crystal System: Orthorhombic
Color: Orange to brownish yellow
Form: Crystals prismatic, vertically striated, capillary or acicular in radiating clusters; massive radial, ocherous, colloform, compact
Relief:
Birefringence:
2V:
Nalpha or Nord.:
NBeta or Nextr.:
NGamma:
Optical Sign:
Orientation:
Pleochroism:
Twinning:
Cleavage:
Extinction:
Alteration:
Features: Polymorphous with akaganeite, feroxyhyte, and lepidocrocite
Occurrence: Secondary iron oxide, usually the product of weathering

Physical Properties

Name: Graphite
Group: Graphite
Formula: C
Crystal System: Hexagonal and trigonal
Color: Iron-black to steel-grey
Opacity: Opaque
Luster: Metallic, dull, earthy
Streak: Black
SGLow: 2.09
SGHigh: 2.23
HardnessLow: 1
HardnessHigh: 2
Cleavage: 1
Direction: {0001} perfect
Habit: Crystals thin tabular, hexagonal; massive, fine to coarse foliated; radiate aggregates, columnar, granular, scaly, earthy
Fracture: Flexible, inelastic; greasy feel; sectile
Other: A high grade metamorphic mineral characteristic of metamorphic gneisses, schists and limestones
Comments: Polymorphous with chaoite, diamond and lonsdaleite

Optical Properties

Name: Graphite
Group: Graphite
Formula: C
Crystal System: Hexagonal and trigonal
Color: Black
Form: Crystals thin tabular, hexagonal; massive, fine to coarse foliated; radiate aggregates, columnar, granular, scaly, earthy
Relief:
Birefringence:
2V:
Nalpha or Nord.:
NBeta or Nextr.:
NGamma:
Optical Sign:
Orientation:
Pleochroism:
Twinning:
Cleavage:
Extinction:
Alteration:
Features: Polymorphous with chaoite, diamond and lonsdaleite
Occurrence: A high grade metamorphic mineral characteristic of metamorphic gneisses, schists and limestones

Physical Properties

Name: Grossular (Grossularite)

Group: Garnet

Formula: Ca3Al2(SiO4)3

Crystal System: Cubic

Color: Colorless, white, grey, yellow to green, brown, pink, red, black

Opacity: Transparent to nearly opaque

Luster: Vitreous to resinous

Streak: White

SGLow: 3.4

SGHigh: 3.6

HardnessLow: 6.5

HardnessHigh: 7

Cleavage: 1

Direction: {110} parting

Habit: Crystals dodecahedrons or trapezohedrons; massive, compact; granular; embedded grains

Fracture: Conchoidal to uneven; brittle

Other: Predominantly found in metamorphic rocks, especially contact metamorphic zones

Comments: Series with andradite with hibschite and katoite, and with uvarovite

Optical Properties

Name: Grossular (Grossularite)

Group: Garnet

Formula: Ca3Al2(SiO4)3

Crystal System: Isometric, cubic

Color: Colorless to pale red, pale brown to brown, greenish

Form: Euhedral dodecahedrons in six sided trapezohedrons in eigth sided cross sections. Plygonal grains and aggregates.

Relief: V.high, n>balsam

Birefringence: Isotropic, but may show weak birefringence

2V:

Nalpha or Nord.: 1.739-1.763

NBeta or Nextr.:

NGamma:

Optical Sign:

Orientation:

Pleochroism:

Twinning:

Cleavage: Parting parallel to{110}. Irregular fractures

Extinction:

Alteration:

Features: Similar to spinel which is octohedral. Determination of RI with differentiate garnets.

Occurrence: Predominantly found in metamorphic rocks, especially contact metamorphic zones

Physical Properties

Name: Grunerite

Group: Amphibole

Formula: (Fe,Mg)7Si8O22(OH)2

Crystal System: Monoclinic

Color: Ash grey, dark green, brown

Opacity: Translucent to nearly opaque

Luster: Silky

Streak:

SGLow: 3.44

SGHigh: 3.60

HardnessLow: 5

HardnessHigh: 6

Cleavage: 1

Direction: {110} good

Habit: Fibrous or fibrolamellar, often in radiating clusters

Fracture:

Other: Chiefly within metamorphic rocks, particularly banded iron formations and mica schists

Comments: Series with magnesiocummingtonite and cummingtonite

Optical Properties

Name: Grunerite

Group: Amphibole

Formula: (Fe,Mg)7Si8O22(OH)2

Crystal System: Monoclinic

Color: Neutral

Form: Fibrous to columnar aggregates, sometimes asbestiform. Cross sections are rhombic

Relief: High, n>balsam

Birefringence: Strong, 0.042-0.054. Sections with parallel extinction are first order.

2V: 79"-86"

Nalpha or Nord.: 1.657-1.663

NBeta or Nextr.: 1.684-1.697

NGamma: 1.699-1.717

Optical Sign: Biaxial negative

Orientation: Length slow, r>v weak

Pleochroism:

Twinning: Polysynthetic twinning characteristic {100}

Cleavage: In two directions at 56" and 124".

Extinction: Oblique 10"-15" in longitudinal sections

Alteration:

Features: Series with magnesiocummingtonite and cummingtonite. Smaller extiction angle than cummingtonite. RI higher than tremolite. Anthophyllite has parallel extinction.

Occurrence: Chiefly within metamorphic rocks, particularly banded iron formations and mica schists

Physical Properties

Name: Gypsum
Group: Evaporite
Formula: CaSO4+2H2O
Crystal System: Monoclinic
Color: Colorless, white, grey, yellowish, greenish, reddish, brownish
Opacity: Transparent
Luster: Subvitreous, pearly on cleavage
Streak: White
SGLow: 2.32
SGHigh: 2.32
HardnessLow: 2
HardnessHigh: 2
Cleavage: 3
Direction: {010} perfect, {100} and {011} distinct
Habit: Crystals thin to thick tabular, diamond-shaped, short to long prismatic; acicular; massive, granular, concretionary
Fracture: Splintery; flexible, not elastic
Other: The main constituent of gypsum rock, formed by hydration of anhydrite
Comments: Sometimes fluoresces and phosphoresces greenish white

Optical Properties

Name: Gypsum
Group: Evaporite
Formula: CaSO4+2H2O
Crystal System: Monoclinic
Color: Colorless
Form: Anhedral to subhedral aggregates, fibrous
Relief: Low, n>balsam
Birefringence: Weak, 0.009
2V: 58"
Nalpha or Nord.: 1.520
NBeta or Nextr.: 1.522
NGamma: 1.529
Optical Sign: Biaxial positive
Orientation: Claevage parallel to slow and fast rays
Pleochroism:
Twinning: Polysynthetic
Cleavage: Three perfect {010}, imperfect {100} and {1_11}
Extinction: Parallel to cleavage
Alteration:
Features: Sometimes fluoresces and phosphoresces greenish white. Weak birefringence and low relief compared to anhydrite.
Occurrence: The main constituent of gypsum rock, formed by hydration of anhydrite

Physical Properties

Name: Halite

Group: Evaporite

Formula: NaCl

Crystal System: Isometric, cubic

Color: Colorless, white,yellow, orange, reddish, purple, blue

Opacity: Transparent to translucent

Luster: Vitreous

Streak: White

SGLow: 2.17

SGHigh: 2.17

HardnessLow: 2

HardnessHigh: 2

Cleavage: 3

Direction: {001} perfect

Habit: Crystals cubic, rarely octahedral, often hopper-shaped or cavernous; massive, compact or granular; columnar; stalactitic

Fracture: Conchoidal; brittle

Other: Characteristic of sedimentary evaporite deposits

Comments: Sometimes fluoresces orange, reddish, or greenish in UV light due to inclusions of organic or inorganic impurities; soluble in water

Optical Properties

Name: Halite

Group: Evaporite

Formula: NaCl

Crystal System: Isometric, cubic

Color: Colorless

Form: Usually anhedral crystals

Relief: Very low, about the same as balsam

Birefringence: Nil, dark between crossed polars

2V:

Nalpha or Nord.: 1.544

NBeta or Nextr.:

NGamma:

Optical Sign:

Orientation:

Pleochroism:

Twinning:

Cleavage: Perfect, three directions

Extinction:

Alteration: Soluble in water

Features: Low relief, cubic form and solubility

Occurrence: Characteristic of sedimentary evaporite deposits

Physical Properties

Name: Halloysite
Group: Kaolinite, Clay
Formula: Al2Si2O5(OH)4
Crystal System: Monoclinic
Color: Colorless, white; may be tinted yellowish, brownish, reddish, or bluish by impurities
Opacity: Transparent to translucent
Luster: Pearly to dull earthy
Streak: White
SGLow: 2
SGHigh: 2.2
HardnessLow: 2
HardnessHigh: 2.5
Cleavage: 0
Direction:
Habit: Tubular (on microscopic scale); massive, compact or mealy
Fracture: Earthy
Other: Associated with other clay minerals in clay beds and altered limestones
Comments: Polymorphous with dickite, kaolinite and nacrite

Optical Properties

Name: Halloysite
Group: Kaolinite, Clay
Formula: Al2Si2O5(OH)4
Crystal System: Monoclinic
Color: Colorless
Form: Fine grained and colloform masses that may show shatter cracks, tubular (on microscopic scale)
Relief: Low, n>balsam
Birefringence: Very weak, almost isotropic
2V:
Nalpha or Nord.: 1.549-1.561
NBeta or Nextr.:
NGamma:
Optical Sign:
Orientation:
Pleochroism:
Twinning:
Cleavage:
Extinction:
Alteration:
Features: Similar to other clay minerals but weak birefringence, RI=balsam and shatter cracks.
Occurrence: Associated with other clay minerals in clay beds and altered limestones

Physical Properties

Name: Hauyne
Group: Sodalite, Feldspathoid
Formula: $(Na,Ca)_{4-8}Al_6Si_6(O,S)_{24}(SO_4,Cl)_{1-2}$
Crystal System: Cubic
Color: Blue, white, grey, green, yellow, red
Opacity: Transparent to translucent
Luster: Vitreous to greasy
Streak: White
SGLow: 2.44
SGHigh: 2.50
HardnessLow: 5.5
HardnessHigh: 6
Cleavage: 1
Direction: {110} distinct
Habit: Crystals dodecahedral or octahedral; rounded grains
Fracture: Conchoidal to uneven; brittle
Other: Occurs within alkali igneous rocks
Comments:

Optical Properties

Name: Hauyne
Group: Sodalite, Feldspathoid
Formula: $(Na,Ca)_{4-8}Al_6Si_6(O,S)_{24}(SO_4,Cl)_{1-2}$
Crystal System: Isometric, cubic
Color: Colorless, grey, pale blue, bluish green
Form: Euhedral octohedrons and dodecahedrons. Crystal aggregates and anhedral crystals
Relief: Low, n<balsam
Birefringence: Isotropic, but may show weak birefringence
2V:
Nalpha or Nord.: 1.496-1.510
NBeta or Nextr.:
NGamma:
Optical Sign:
Orientation:
Pleochroism:
Twinning:
Cleavage: May show one imperfect cleavage
Extinction: Isotropic
Alteration:
Features: Sulphide (mostly pyrite) bearing variety known as lapis lazuli. Crystals and RI somewhat distinctive
Occurrence: Occurs within alkali igneous rocks, such as phonolite, and within contact metamorphic gneisses in the form of lapis lazuli.

Physical Properties

Name: Hedenbergite

Group: Pyroxene

Formula: CaFeSi2O6

Crystal System: Monoclinic

Color: Brownish green, greyish green, dark green, greyish black, black

Opacity: Translucent to nearly opaque

Luster: Vitreous to resinous or dull

Streak: White, greyish

SGLow: 3.50

SGHigh: 3.56

HardnessLow: 6

HardnessHigh: 6

Cleavage: 3

Direction: {110} good, {100} and {010} parting

Habit: Crystals short prismatic; massive, lamellar

Fracture: Conchoidal to uneven; brittle

Other: In contact metamorphic zones and skarns

Comments: Series with diopside and with johannsenite

Optical Properties

Name: Hedenbergite

Group: Pyroxene

Formula: CaFeSi2O6

Crystal System: Monoclinic

Color: Neutral to greenish

Form: Usually in columnar aggregates

Relief: V. high, n>balsam

Birefringence: Moderate, 0.018-0.019

2V: 60"

Nalpha or Nord.: 1.732-1.739

NBeta or Nextr.: 1.7737-1.745

NGamma: 1.751-1.757

Optical Sign: Biaxial positive

Orientation: Length fast, r>v weak

Pleochroism:

Twinning:

Cleavage: In two directions at 87" and 93"

Extinction: Oblique in longitudinal sections, 42", parallel in cross sections

Alteration:

Features: Higher RI than diopside and augite. Occurrence distinctive.

Occurrence: In contact metamorphic zones and skarns

Physical Properties

Name: Hematite
Group: Hematite
Formula: Fe2O3
Crystal System: Trigonal
Color: Steel-grey to iron-black, thin fragments deep blood red
Opacity: Opaque
Luster: Metallic, submetallic, dull
Streak: Deep red or brownish red
SGLow: 5.26
SGHigh: 5.26
HardnessLow: 5
HardnessHigh: 6
Cleavage: 2
Direction: {0001} and {10-11} parting due to twinning
Habit: Crystals thin to thick tabular, rhombohedral, pyramidal, or prismatic; tabular crystals may form rosettes; massive
Fracture: Subconchoidal to uneven; brittle
Other: Characteristic of banded iron formations and as a secondary mineral in many rock types
Comments: Dimorphous with maghemite

Optical Properties

Name: Hematite
Group: Hematite
Formula: Fe2O3
Crystal System: Trigonal
Color: Deep red or brownish red
Form: Crystals thin to thick tabular, rhombohedral, pyramidal, or prismatic; tabular crystals may form rosettes; massive
Relief:
Birefringence:
2V:
Nalpha or Nord.:
NBeta or Nextr.:
NGamma:
Optical Sign:
Orientation:
Pleochroism:
Twinning:
Cleavage:
Extinction:
Alteration:
Features: Dimorphous with maghemite
Occurrence: Characteristic of banded iron formations and as a secondary mineral in many rock types

Physical Properties

Name: Heulandite
Group: Zeolite
Formula: (Na,Ca)2-3Al3(Al,Si)2Si13O36+12H2O
Crystal System: Monoclinic
Color: Colorless, white, grey, yellow, pink, red, brown
Opacity: Transparent to translucent
Luster: Vitreous, pearly on {010}
Streak: Colorless
SGLow: 2.1
SGHigh: 2.2
HardnessLow: 3.5
HardnessHigh: 4
Cleavage: 1
Direction: {010} perfect
Habit: Crystals trapezoidal, tabular; subparallel aggregates; massive, granular
Fracture: Uneven; brittle
Other: In cavities and veins within lavas, particularly basalts
Comments:

Optical Properties

Name: Heulandite
Group: Zeolite
Formula: (Na,Ca)2-3Al3(Al,Si)2Si13O36+12H2O
Crystal System: Monoclinic
Color: Colorless
Form: Tabular subhedral to euhedral crystals
Relief: Low, n<balsam
Birefringence: Weak, 0.007
2V: 0"-48"
Nalpha or Nord.: 1.496-1.499
NBeta or Nextr.: 1.497-1.501
NGamma: 1.501-1.505
Optical Sign: Biaxial positive
Orientation: Length fast, r<v
Pleochroism:
Twinning:
Cleavage: One perfect {010}
Extinction: Parallel to cleavage
Alteration:
Features: Biaxial positive, tabular crystals and parallel extinction.
Occurrence: In cavities and veins within lavas, particularly basalts

Physical Properties

Name: Hornblende

Group: Amphibole

Formula: $(Na,K)0-1(Ca_2(Mg,Fe^{2+},Fe^{3+},Al)5[Si6-7Al2-1)22](OH,$

Crystal System: Monoclinic

Color: Green, black, dark green, yellow-brown, brown

Opacity: Translucent

Luster: Vitreous

Streak: Colorless

SGLow: 3.02

SGHigh: 3.45

HardnessLow: 5

HardnessHigh: 6

Cleavage: 2

Direction: {110}, {100} good

Habit: Prismatic

Fracture: Uneven, brittle

Other: Most widespread of the amphiboles. Widespread within igneous and metamorphic rocks

Comments: A subgroup name in the amphibole group

Optical Properties

Name: Hornblende

Group: Amphibole

Formula: $(Na,K)0-1(Ca_2(Mg,Fe^{2+},Fe^{3+},Al)5[Si6-7Al2-1)22](OH,$

Crystal System: Monoclinic

Color: Green or brown in various shades

Form: Prismatic with pseudo-hexagonal cross sections

Relief: High, n>balsam

Birefringence: Moderate, 0.019-0.026

2V: 52"-85"

Nalpha or Nord.: 1.614-1.675

NBeta or Nextr.: 1.618-1.691

NGamma: 1.633-1.701

Optical Sign: Biaxial negative

Orientation: r<v weak

Pleochroism: Strong pleochroism from yellow green-dark green, pale brown-dark green, yellow green - brown, greenish brown- red brown

Twinning: Simple twinning along {100} common

Cleavage: In two directions {110} at 56" and 124"

Extinction: Oblique 12" to 30" in longitudinal sections, symmetrical in cross sections

Alteration: Alters to chlorite and epidote

Features: Differs from pyroxenes (augite) in cleavgae, pleochroism and extinction angle which are distinctive.

Occurrence: Most widespread of the amphiboles. Widespread within igneous and metamorphic rocks. Also within detrital sediments.

Physical Properties

Name: Humite
Group: Humite
Formula: (Mg,Fe)7(SiO4)3(F,OH)2
Crystal System: Orthorhombic
Color: White, yellow, dark orange, brown
Opacity: Transparent to translucent
Luster: Vitreous
Streak:
SGLow: 3.24
SGHigh: 3.24
HardnessLow: 6
HardnessHigh: 6
Cleavage: 1
Direction: {100} indistinct
Habit: Crystals varied habit, highly modified, small
Fracture: Subconchoidal to uneven; brittle
Other: Occurs in metamorphosed limestones
Comments:

Optical Properties

Name: Humite
Group: Humite
Formula: (Mg,Fe)7(SiO4)3(F,OH)2
Crystal System: Orthorhombic
Color:
Form:
Relief:
Birefringence:
2V:
Nalpha or Nord.:
NBeta or Nextr.:
NGamma:
Optical Sign:
Orientation:
Pleochroism:
Twinning:
Cleavage:
Extinction:
Alteration:
Features:
Occurrence: Occurs in metamorphosed limestones

Physical Properties

Name: Hypersthene

Group: Pyroxene

Formula: (MgFe)SiO3

Crystal System: Orthorhombic

Color: Greyish, greenish, brown to bronze

Opacity: Transparent to transluscent

Luster: Bronze luster

Streak: Brownish grey, greyish white

SGLow: 3.4

SGHigh: 3.9

HardnessLow: 5

HardnessHigh: 6

Cleavage: 3

Direction: {210} perfect, {100}, {010} indistinct

Habit: Coarsely crystalline and foliated cleavage masses

Fracture: Uneven

Other: Occurs within igneous rocks, especially charnockite, norite, hypersthene gabbro, tholiitic basalts

Comments: A member of the enstatite-ferrosilite series

Optical Properties

Name: Hypersthene

Group: Pyroxene

Formula: (MgFe)SiO3

Crystal System: Orthorhombic

Color: Neutral to pale green or pale red

Form: Subhedral prismatic crystals, nearly square cross sections. Inclusions produce shiller structure

Relief: High, n>balsam

Birefringence: Weak, 0.010-0.016, first order yellow to red maximum

2V: 63"-90"

Nalpha or Nord.: 1.673-1.715

NBeta or Nextr.: 1.678-1.728

NGamma: 1.683-1.731

Optical Sign: Biaxial negative

Orientation: Length slow, r>v weak

Pleochroism: Pleochroic from greenish-pale red

Twinning:

Cleavage: Three distinct {110}, sometimes {010} and {100}

Extinction: Parallel

Alteration:

Features: Pleochroism distinctive

Occurrence: Occurs within igneous rocks, especially charnockite, norite, hypersthene gabbro, tholiitic basalts

Optical Properties

Name: Iddingsite

Group: Garnet

Formula: MgO.Fe2O3.3SiO2.4H2O

Crystal System: Orthorhombic

Color: Brown

Form: Always as aprtial or complete pseudomorphs after olivine

Relief: High, n>balsam

Birefringence: Strong, 0.038-0.044

2V: 25"-60"

Nalpha or Nord.: 1.674-1.730

NBeta or Nextr.: 1.715-1.763

NGamma: 1.718-1.768

Optical Sign: Biaxial positive or negative

Orientation: r>v or r<v strong

Pleochroism: Weak to moderate pleochroism

Twinning:

Cleavage: Three directions {100}, {001} and {010} at right angles

Extinction: Parallel to cleavage

Alteration: Alters to limonite and hydrous iron oxides

Features: Redish-brown color and cleavage lamellar structure and occurrence are distinctive.

Occurrence: Occurs in basalts and basalt porphyries as an hydrothermal alteration product of olivine.

Physical Properties

Name: Illite
Group: Illite, Clay
Formula: (K,H3O)(Al,Mg,Fe)2(Si,Al)4O10[(OH)2,H2O]
Crystal System: Orthorhombic
Color: White and pale colors
Opacity: Opaque
Luster: Dull
Streak:
SGLow: 2.6
SGHigh: 2.9
HardnessLow: 1
HardnessHigh: 2
Cleavage: 1
Direction: {001} perfect
Habit: Massive, extremely fine-grained
Fracture:
Other: A weathering product widespread in soils and in clay beds. Also common as a diagenetic clay mineral in sedimentary beds
Comments: A group of mica-clay minerals, not a distinct species

Optical Properties

Name: Illite
Group: Illite, Clay
Formula: (K,H3O)(Al,Mg,Fe)2(Si,Al)4O10[(OH)2,H2O]
Crystal System: Orthorhombic
Color: Colorless to yellowish-brown
Form: Irregular matted flakes that may be intergrown with montmorillonite or kaolinite
Relief: Low, n>balsam
Birefringence: Strong, 0.030-0.035, but might not show above 2nd order due to small size
2V: Small
Nalpha or Nord.: 1.535-1.570
NBeta or Nextr.:
NGamma: 1.565-1.605
Optical Sign: Biaxial negative
Orientation:
Pleochroism:
Twinning:
Cleavage: One perfect {001}
Extinction:
Alteration:
Features: Microcrystalline matted flakes may be distinctive. XRD may be necessary.
Occurrence: A weathering product widespread in soils and in clay beds. Also common as a diagenetic clay mineral in sedimentary beds

Optical Properties

Name: Ilmenite

Group: Oxides

Formula: $FeTiO_3$

Crystal System: Hexagonal

Color: Blue-grey, black

Form: Disseminated tabular crystals. Skeleton crystals occur. Also in irregular grains and masses.

Relief: V. high

Birefringence: Opaque, metallic luster in reflected light

2V:

Nalpha or Nord.:

NBeta or Nextr.:

NGamma:

Optical Sign:

Orientation:

Pleochroism:

Twinning:

Cleavage:

Extinction:

Alteration:

Features: Opaque, metallic luster, tabular crystals.

Occurrence: Widely distributed in igneous rocks , especially dolerites. Important titanium mineral in sedimentary heavy mineral sand deposits.

Physical Properties

Name: Jadeite
Group: Pyroxene
Formula: NaAlSi2O6
Crystal System: Monoclinic
Color: Crystals colorless with green tips; massive in many colors
Opacity: Transparent to translucent
Luster: Vitreous, massive is greasy to vitreous
Streak: Colorless
SGLow: 3.25
SGHigh: 3.25
HardnessLow: 6
HardnessHigh: 6
Cleavage: 1
Direction: {110} good
Habit: Crystals elongated prismatic, small, rare; massive fine granular to coarse granular; as small to large alluvial boulders
Fracture: Splintery, uneven
Other: Characteristic of high grade metamorphic rocks
Comments:

Optical Properties

Name: Jadeite
Group: Pyroxene
Formula: NaAlSi2O6
Crystal System: Monoclinic
Color: Colorless to green
Form: Granular to columnar or fibrous. Fine to coarse grained, euhedral crystals rare.
Relief: High, n>balsam
Birefringence: Moderate, 0.012-0.023
2V: 70"-75"
Nalpha or Nord.: 1.655-1.666
NBeta or Nextr.: 1.659-1.674
NGamma: 1.667-1.688
Optical Sign: Biaxial positive
Orientation: Length slow, r<v
Pleochroism: Deeply colored varieties are pleochroic
Twinning: Simple twins along {100} infrequent
Cleavage: In two directions at 87" and 93", parallel in longitudinal sections
Extinction: Oblique 30"-44" in longitudinal sections
Alteration: Alters to tremolite/actinolite
Features: Higher extinction angle and RI than nephrite, smaller extinction angle and columnar habit distinguishes from diopside
Occurrence: Characteristic of high grade metamorphic rocks

Physical Properties

Name: Jarosite

Group: Alunite

Formula: KFe3(SO4)2(OH)6

Crystal System: Trigonal

Color: Pale yellow to yellowish brown to brown

Opacity: Translucent

Luster: Vitreous to resinous

Streak: Pale yellow

SGLow: 2.9

SGHigh: 3.26

HardnessLow: 2.5

HardnessHigh: 3.5

Cleavage: 1

Direction: {0001} distinct

Habit: Crystals tabular or pseudocubic, microscopic to minute as crusts and coatings; massive, granular; fibrous; earthy

Fracture: Conchoidal to uneven; brittle

Other: A secondary mineral that occurs as thin bands in sedimentary shales and sometimes in volcanic rocks

Comments:

Optical Properties

Name: Jarosite

Group: Alunite

Formula: KFe3(SO4)2(OH)6

Crystal System: Hexagonal

Color: Colorless to brown

Form: Aggregates of fine grains, sometimes euhedral

Relief: V. high, n>balsam

Birefringence: Extreme, 0.105

2V:

Nalpha or Nord.: 1.715

NBeta or Nextr.: 1.820

NGamma:

Optical Sign: Uniaxial negative

Orientation:

Pleochroism:

Twinning:

Cleavage: One direction, {0001}

Extinction: Parallel or symmetrical

Alteration: Alters to limonite

Features: Crystals, birefringence, cleavage

Occurrence: A secondary mineral that occurs as thin bands in sedimentary shales and sometimes in volcanic rocks

Physical Properties

Name: Kaolinite
Group: Kaolinite, Clay
Formula: Al2Si2O5(OH)4
Crystal System: Triclinic
Color: Colorless, white, yellowish pink, reddish, bluish
Opacity: Transparent to translucent
Luster: Pearly to dull earthy
Streak: White
SGLow: 2.6
SGHigh: 2.63
HardnessLow: 2
HardnessHigh: 2.5
Cleavage: 1
Direction: {001} perfect
Habit: Thin hexagonal platelets or scales; elongated platelets or curved laths; massive compact, friable, mealy
Fracture: Scales flexible, inelastic, often plastic when moist
Other: A weathering product of igneous and metamorphic rocks and in clay beds. Also common as a diagenetic clay mineral in sedimentary beds
Comments: Polymorphous with dickite, halloysite and nacrite

Optical Properties

Name: Kaolinite
Group: Kaolinite, Clay
Formula: Al2Si2O5(OH)4
Crystal System: Triclinic
Color: Colorless to pale yellow
Form: Microcrystalline aggregates of scales and plates with six sided outline, in veins replacing feldspars
Relief: Low, n>balsam
Birefringence: Weak, 0.005
2V: Variable
Nalpha or Nord.: 1.561
NBeta or Nextr.: 1.565
NGamma: 1.566
Optical Sign: Biaxial negative
Orientation: Length slow
Pleochroism:
Twinning:
Cleavage: One perfect {001}
Extinction: Almost parallel 1'-3.5" along base
Alteration:
Features: Low birefringence, aggregates of hexagonal scales and plates
Occurrence: A weathering product of igneous and metamorphic rocks and in clay beds. Also common as a diagenetic clay mineral in sedimentary beds

Physical Properties

Name: Kernite
Group: Borate
Formula: Na2B4O6(OH)2+3H2O
Crystal System: Monoclinic
Color: Colorless; white when surface altered
Opacity: Transparent
Luster: Dull to vitreous, silky or pearly on cleavage
Streak:
SGLow: 1.91
SGHigh: 1.91
HardnessLow: 2.5
HardnessHigh: 3
Cleavage: 3
Direction: {100} and {001} perfect; {-201} distinct
Habit: Crystals nearly equant; massive, fibrous structure, cleavable
Fracture: Brittle, splintery
Other: Occurs within beds in desert environments
Comments: Opaque when surface altered

Optical Properties

Name: Kernite
Group: Borate
Formula: Na2B4O6(OH)2+3H2O
Crystal System: Monoclinic
Color: Colorless
Form: Large crystals upto 1m thick
Relief: Low to moderate, n< balsam
Birefringence: Moderate to strong, 0.034
2V: 80"
Nalpha or Nord.: 1.454
NBeta or Nextr.: 1.472
NGamma: 1.488
Optical Sign: Biaxial negative
Orientation:
Pleochroism:
Twinning:
Cleavage: Three {100} perfect, {001} good, {201} fair
Extinction:
Alteration:
Features: Similar to gypsum, but soluble in warm water
Occurrence: Occurs within beds in desert environments

Physical Properties

Name: Kyanite
Group: Kyanite
Formula: Al2SiO5
Crystal System: Triclinic
Color: Blue, colorless, white, grey, green, yellow, pink, nearly black
Opacity: Transparent to translucent
Luster: Vitreous to pearly
Streak: Colorless
SGLow: 3.53
SGHigh: 3.67
HardnessLow: 4
HardnessHigh: 7.5
Cleavage: 3
Direction: {100} perfect, {010} distinct, {001} parting
Habit: Crystals long bladed, elongated parallel to c-axis; often bent or twisted; massive, bladed to fibrous
Fracture: Somewhat flexible
Other: Widespread within metamorphic rocks
Comments: Trimorphous with andalusite and sillimanite

Optical Properties

Name: Kyanite
Group: Sillimanite
Formula: Al2SiO5
Crystal System: Triclinic
Color: Colorless to pale blue
Form: Broad, tabular plates parallel to {100} and narrow sections parallel to {010}. Often bent crystals
Relief: High, n>balsam
Birefringence: Moderate, 0.016
2V: 82"
Nalpha or Nord.: 1.712
NBeta or Nextr.: 1.720
NGamma: 1.728
Optical Sign: Biaxial negative
Orientation: Length slow, r>v
Pleochroism: May be p[leochroic in thick sections
Twinning: Frequent along {100} and {001}
Cleavage: Three perfect parallel to {100}, distinct parallel to {010} at right angles. Parting at 85" to length.
Extinction: Oblique 30" in longitudinal sections, parallel or nearly so in cross sections
Alteration:
Features: Extinction angle of 30" and biaxial figure are distinctive
Occurrence: Widespread within metamorphic rocks, including schists, gneisses, some eclogites and pegmatites.

Physical Properties

Name: Labradorite
Group: Feldspar
Formula: (Na,Ca)(Al,Si)4O8
Crystal System: Triclinic
Color: Colorless, white, grey; frequently shows play of colors
Opacity: Transparent to translucent
Luster: Vitreous
Streak: White
SGLow: 2.69
SGHigh: 2.72
HardnessLow: 6
HardnessHigh: 6.5
Cleavage: 3
Direction: {001} perfect, {010} and {110} nearly perfect
Habit: Crystals tabular; massive, cleavable, granular or compact
Fracture: Conchoidal to uneven; brittle
Other: Most abundant in mafic igneous rocks. , such as basalt, gabbro and olivine gabbro. Widespread occurrence
Comments:

Optical Properties

Name: Labradorite
Group: Feldspar
Formula: (Na,Ca)(Al,Si)4O8 An50-70%
Crystal System: Triclinic
Color: Colorless
Form: Anhedral to euhedral triclinic pinacoids
Relief: Low, n>balsam
Birefringence: 2.69
2V: 76"-90"
Nalpha or Nord.: 1.554-1.564
NBeta or Nextr.: 1.558-1.569
NGamma: 1.562-1.573
Optical Sign: Biaxial positive
Orientation: r<v
Pleochroism:
Twinning: According to albite, carlsbad and pericline laws, as in albite
Cleavage: Four perfect {001}, distinct {010}, imperfect ({110} and {11_0}
Extinction: Oblique 27.5" to 39" in albite twins. {001} cleavage sections minus 7"- minus 16", on {010} minus 16" - minus 29"
Alteration: Alters to clay
Features: Maximum extinction angle of albite twins and RI distinctive
Occurrence: Most abundant in mafic igneous rocks, such as basalt, gabbro and olivine gabbro. Widespread occurrence

Optical Properties

Name: Lamprobolite (Basaltic Hornblende)

Group: Amphibole

Formula: Ca,Mg,Fe,Al Silicate

Crystal System: Monoclinic

Color: Yellow to brown, with opaque borders

Form: Euhedral pseudohexagonal crystals, short prismatic

Relief: High, n>balsam

Birefringence: Strong to v.strong, 0.026-0.072

2V: 64"-80"

Nalpha or Nord.: 1.670-1.692

NBeta or Nextr.: 1.683-1.730

NGamma: 1.693-1.760

Optical Sign: Biaxial negative

Orientation: L:ength slow, r<v

Pleochroism: Strong pleochroism Ffrom light yellow-brown-dark red brown

Twinning: Simple twins not distinct

Cleavage: In two directions {110} at 56" and 124"

Extinction: Oblique to parallel 0"-12" in longitudinal sections

Alteration:

Features: Dark staining along rims distinctive. Smaler extinction angle and strong birefringence distinguishes from ordinary brown hornblende.

Occurrence: Phenocrysts in igneous volcanic rocks, including andesites, basalts, basanites tephrites and tuffs.

Physical Properties

Name: Laumontite (Zeolite)
Group: Zeolite
Formula: $CaAl_2Si_4O_{12}+4H_2O$
Crystal System: Monoclinic
Color: White, grey, yellowish, pink, brownish
Opacity: Transparent to translucent
Luster: Vitreous to pearly
Streak: Colorless
SGLow: 2.20
SGHigh: 2.41
HardnessLow: 3
HardnessHigh: 4
Cleavage: 2
Direction: {010} and {110} perfect
Habit: Crystals square prisms with steep oblique terminations; fibrous, columnar, radiating and divergent
Fracture: Uneven
Other: In veins and cavities of igneous rocks
Comments: Readily loses water on exposure and crumbles

Optical Properties

Name: Laumontite (Zeolite)
Group: Zeolite
Formula: $CaAl_2Si_4O_{12}+4H_2O$
Crystal System: Monoclinic
Color: Colorless
Form: Prismatic or fibrous
Relief: Low to moderate
Birefringence: Weak, 0.1012
2V: 26"-47"
Nalpha or Nord.: 1.502-1.514
NBeta or Nextr.: 1.512-1.522
NGamma: 1.514-1.525
Optical Sign: Biaxial negative
Orientation: Length slow, r<v strong
Pleochroism:
Twinning:
Cleavage: Two perfect {010} and {110}
Extinction: Usually oblique
Alteration:
Features: Length slow crystals
Occurrence: In veins and cavities of igneous rocks

Physical Properties

Name: Lawsonite
Group: Epidote
Formula: CaAl2Si2O7(OH)2+H2O
Crystal System: Orthorhombic
Color: Colorless, white, grey, blue, pinkish
Opacity: Translucent
Luster: Vitreous to greasy
Streak: White
SGLow: 3.05
SGHigh: 3.12
HardnessLow: 6
HardnessHigh: 6
Cleavage: 3
Direction: {100} and {010} perfect, {101} imperfect
Habit: Crystals prismatic or tabular; subhedral tablets; massive, granular
Fracture:
Other: Occurs within high grade metamorphic rocks, especialy glaucophane schists
Comments: Dimorphous with partheite

Optical Properties

Name: Lawsonite
Group: Epidote
Formula: CaAl2Si2O7(OH)2+H2O
Crystal System: Orthorhombic
Color: Colorless
Form: Euhedral crystals with rhombic or rectangular outlines
Relief: High, n>balsam
Birefringence: Moderate, 0.019
2V: 84"
Nalpha or Nord.: 1.665
NBeta or Nextr.: 1.674
NGamma: 1.684
Optical Sign: Biaxial positive
Orientation: Length slow along long direction of rhombic cross section, r>c strong
Pleochroism: May be pleochroic in thick sections
Twinning: Polysynthetic along {110} frequent
Cleavage: Three good {010} and {001}, fair {110}, rhombic cleavage.
Extinction: Parallel in longitudinal sections, symmetrical in cross sections
Alteration:
Features: Euhedral rhombic outlines and cleavage distinguishes from clinozoisite
Occurrence: Occurs within high grade metamorphic rocks, especialy glaucophane schists. Also metamorphic gabbros and diorites.

Physical Properties

Name: Lazulite

Group: Apatite, Phosphates

Formula: MgAl2(PO4)2(OH)2

Crystal System: Monoclinic

Color: Deep azure blue to light blue, bluish green

Opacity: Transparent to opaque

Luster: Vitreous to dull

Streak: White

SGLow: 3.10

SGHigh: 3.10

HardnessLow: 5.5

HardnessHigh: 6

Cleavage: 2

Direction: {110} indistinct to good; {101} indistinct

Habit: Crystals acute pyramidal, sometimes tabular; massive, compact to granular

Fracture: Uneven to splintery; brittle

Other: Characteristic of metamorphic rocks, especially quartzites and in quartz veins

Comments: Series with scorzalite

Optical Properties

Name: Lazulite

Group: Apatite, Phosphates

Formula: MgAl2(PO4)2(OH)2

Crystal System: Monoclinic

Color: Blue to colorless

Form: Anhedral crystals, sometimes euhedral bipyrimidal crystals

Relief: High, n>balsam

Birefringence: Strong, 0.036-0.038

2V: 69"

Nalpha or Nord.: 1.603

NBeta or Nextr.: 1.632-1.604

NGamma: 1.639-1.642

Optical Sign: Biaxial negative

Orientation: Length fast

Pleochroism: Some sections pleochroic, colorless-blue

Twinning: Polysynthetic twinning common

Cleavage: One indistinct parallel to {110}

Extinction: Oblique

Alteration:

Features: Blue pleochroism with strong birefringence

Occurrence: Characteristic of metamorphic rocks, especially quartzites and in quartz veins

Physical Properties

Name: Lepidolite
ID: None
Group: Mica
Formula: K(Li,Al)3(Si,Al)4O10(F,OH)2
Crystal System: Monoclinic
Color: Pink to purple, colorless, white, greyish, yellowish
Opacity: Transparent to translucent
Luster: Pearly
Streak: Colorless
SGLow: 2.8
SGHigh: 3.3
HardnessLow: 2.5
HardnessHigh: 3
Cleavage: 3
Direction: {001} perfect, micaceous, {110} and {010} imperfect
Habit: Crystals tabular, pseudohexagonal or hexagonal; thick cleavable masses, coarse to fine scaly; aggregates of books
Fracture: Flexible, inelastic
Other: A lithium mineral found mainly in granite pegmatites with other lithium minerals, spodumene
Comments:

Optical Properties

Name: Lepidolite
Group: Mica
Formula: K(Li,Al)3(Si,Al)4O10(F,OH)2
Crystal System: Monoclinic
Color: Colorless
Form: Short prismatic pseudo-hexagonal and thick tabular crystals
Relief: Fair, n>balsam
Birefringence: Strong, 0.045
2V: 40"
Nalpha or Nord.: 1.560
NBeta or Nextr.: 1.598
NGamma: 1.605
Optical Sign: Biaxial negative
Orientation: Length slow, r>v weak
Pleochroism:
Twinning: according to mica law {110}
Cleavage: One perfect {001}
Extinction: Oblique to parallel 0"-7" along cleavage
Alteration:
Features: Similar to muscovite but larger axial angle. Occurence with other lithium minerals may be distinctive
Occurrence: A lithium mineral found mainly in granite pegmatites with other lithium minerals, spodumene

Physical Properties

Name: Leucite
Group: Feldspathoid
Formula: KAlSi2O6
Crystal System: Tetragonal
Color: Colorless, white, grey
Opacity: Transparent to translucent
Luster: Vitreous
Streak: Colorless
SGLow: 2.47
SGHigh: 2.50
HardnessLow: 5.5
HardnessHigh: 6
Cleavage: 1
Direction: {110} imperfect
Habit: Crystals trapezohedral, faces sometimes finely striated from twinning; disseminated grains; massive, granular
Fracture: Conchoidal; brittle
Other: Occurs mostly as phenocrysts in lavas
Comments:

Optical Properties

Name: Leucite
Group: Feldspathoid
Formula: KAlSi2O6
Crystal System: Tetragonal, pseudo isometric
Color: Colorless
Form: Euhdral showing octohedral sections
Relief: Fair, n<balsam
Birefringence: Weak, 0.001
2V:
Nalpha or Nord.: 1.508
NBeta or Nextr.: 1.509
NGamma:
Optical Sign:
Orientation:
Pleochroism:
Twinning: Complicated polysynthetic twinning in several directions is characteristic
Cleavage: One {110} imperfect
Extinction: Often wavy
Alteration:
Features: Twinning and octohedral sections distinctive
Occurrence: Occurs mostly as phenocrysts in alkali lavas

Optical Properties

Name: Leucoxene

Group: Mineraloid

Formula: FeTiO3

Crystal System: Amorphous

Color: Opaque white

Form: Along borders of titanium minerals and dispersed grains

Relief:

Birefringence: Opaque

2V:

Nalpha or Nord.:

NBeta or Nextr.:

NGamma:

Optical Sign:

Orientation:

Pleochroism:

Twinning:

Cleavage:

Extinction:

Alteration:

Features: Associated with titanium minerals

Occurrence: Alteration product of titanium minerals

Physical Properties

Name: Limonite
Group: Limonite-Goethite
Formula: FeO.OH.nH2O
Crystal System: Amorphous/ Cryptocrystalline
Color: Yellow, brown, brownish-black, orange-brown
Opacity: Opaque
Luster: Non metallic
Streak: Yellowish-brown
SGLow: 2.7
SGHigh: 4.3
HardnessLow: 4
HardnessHigh: 5.5
Cleavage: None
Direction: None
Habit: Massive, earthy, vitreous
Fracture: Uneven, subconchoidal
Other: Secondary iron oxide, usually the product of weathering
Comments: A general term for hydrous iron oxides

Optical Properties

Name: Limonite
Group: Limonite-Goethite
Formula: FeO.OH.nH2O
Crystal System: Amorphous/ Cryptocrystalline
Color: Yellowish-brown
Form: Massive, earthy, vitreous
Relief:
Birefringence:
2V:
Nalpha or Nord.:
NBeta or Nextr.:
NGamma:
Optical Sign:
Orientation:
Pleochroism:
Twinning:
Cleavage:
Extinction:
Alteration:
Features: A general term for hydrous iron oxides
Occurrence: Secondary iron oxide, usually the product of weathering

Physical Properties

Name: Lizardite
Group: Serpentine
Formula: Mg3Si2O5(OH)4
Crystal System: Trigonal and hexagonal
Color: Green, green-blue, white
Opacity: Translucent
Luster: Greasy, waxy, silky
Streak:
SGLow: 2.55
SGHigh: 2.60
HardnessLow: 2.5
HardnessHigh: 2.5
Cleavage: 1
Direction: {0001} perfect
Habit: Massive, fine- to coarse-grained, compact; small scales
Fracture:
Other: Associated with chrysotile in serpentinites
Comments: Polymorphous with antigorite, clinochrysotile, orthochrysotile and parachrysotile; series with nepouite, D=approximate

Optical Properties

Name: Lizardite
Group: Serpentine
Formula: Mg3Si2O5(OH)4
Crystal System: Trigonal and hexagonal
Color: Colorless to pale green
Form: Fine grained, platy
Relief: Low, n>balsam
Birefringence: Moderate, 0.014-0.016
2V:
Nalpha or Nord.: 1.538-1.554
NBeta or Nextr.:
NGamma: 1.546-1.560
Optical Sign: Biaxial negative
Orientation: Length slow
Pleochroism:
Twinning:
Cleavage: One {001} perfect
Extinction: Parallel to elongation
Alteration:
Features: XRD probably required. RI intermediate between antigorite and chrysotile
Occurrence: Associated with chrysotile in serpentinites

Physical Properties

Name: Magnesite

Group: Calcite

Formula: $MgCO_3$

Crystal System: Trigonal

Color: Colorless, white, grey, yellowish to brown

Opacity: Transparent to translucent

Luster: Vitreous, dull

Streak: White

SGLow: 3.0

SGHigh: 3.1

HardnessLow: 3.75

HardnessHigh: 4.25

Cleavage: 1

Direction: {10-11} perfect

Habit: Crystals rhombohedral, uncommon; rarely prismatic, tabular or scalenohedral; massive, compact, chalky, lamellar, fibrous

Fracture: Conchoidal; brittle

Other: Main constituent of metamorphic magnesite rocks. Also occurs in serpentinites

Comments: Series with gaspeite and with siderite

Optical Properties

Name: Magnesite

Group: Calcite

Formula: $MgCO_3$

Crystal System: Hexagonal

Color: Colorless

Form: Anhedral to subhedral crystal aggregates. Euhedral crystals rare.

Relief: High along long direction, low along short direction

Birefringence: Extreme, 0.191-0.199

2V:

Nalpha or Nord.: 1.509-1.527

NBeta or Nextr.: 1.700-1.726

NGamma:

Optical Sign: Uniaxial negative

Orientation:

Pleochroism:

Twinning: None

Cleavage: One perfect rhombohedral along {101_1}

Extinction: Symmetrical to cleavage

Alteration:

Features: Similar to dlomite and calcite. Staining may be required.

Occurrence: Main constituent of metamorphic magnesite rocks. Also occurs in serpentinites

Physical Properties

Name: Magnetite
Group: Spinel
Formula: Fe3O4
Crystal System: Cubic
Color: Iron-black, greyish black
Opacity: Opaque
Luster: Splendent metallic to dull
Streak: Black
SGLow: 5.175
SGHigh: 5.175
HardnessLow: 5.5
HardnessHigh: 6.5
Cleavage: 1
Direction: {111} parting good
Habit: Crystals octahedral, may be highly modified or dodecahedral and striated; massive, compact or fine to coarse granular
Fracture: Subconchoidal to uneven; brittle
Other: Widespread accessory mineral in igneous and metamorphic rocks. Common detrital mineral in sedimentary deposits
Comments: Series with jacobsite and with magnesioferrite

Optical Properties

Name: Magnetite
Group: Spinel
Formula: Fe3O4
Crystal System: Cubic
Color: Black
Form: Crystals octahedral, may be highly modified or dodecahedral and striated; massive, compact or fine to coarse granular
Relief:
Birefringence:
2V:
Nalpha or Nord.:
NBeta or Nextr.:
NGamma:
Optical Sign:
Orientation:
Pleochroism:
Twinning:
Cleavage: One {111} parting good
Extinction:
Alteration:
Features: Series with jacobsite and with magnesioferrite
Occurrence: Widespread accessory mineral in igneous and metamorphic rocks. Common detrital mineral in sedimentary deposits

Physical Properties

Name: Melilite

Group: Melilite

Formula: Al2[C6(COO)6]+18H2O

Crystal System: Tetragonal

Color: Golden brown or brownish, reddish, yellow

Opacity: Transparent to translucent

Luster: Vitreous to resinous

Streak:

SGLow: 1.64

SGHigh: 1.64

HardnessLow: 2

HardnessHigh: 2.5

Cleavage: 1

Direction: {011} indistinct

Habit: Crystals prismatic or pyramidal, small, rare; massive, fine granular; nodular; coatings

Fracture: Conchoidal; somewhat sectile

Other: In alkali igneous rocks

Comments:

Optical Properties

Name: Melilite

Group: Melilite

Formula: Al2[C6(COO)6]+18H2O

Crystal System: Tetragonal

Color: Colorless to pale yellow

Form: Euhedral, tabular, rectangular prisms frequently occur. Peg structure from lines perpendicular to length

Relief: High, n>balsam

Birefringence: Weak, 0.005-0.006

2V:

Nalpha or Nord.: 1.626-1.629

NBeta or Nextr.: 1.632-1.634

NGamma:

Optical Sign: Uniaxial positive

Orientation:

Pleochroism:

Twinning:

Cleavage: One indistinct {001}

Extinction: Parallel

Alteration: Along lines perpendicular to length, giving peg structure, Alters to calcite and zeolites.

Features: Elongated, rectangular sections with peg structure distinctive.

Occurrence: In alkali igneous rocks, such as nepheline and leucite bearing lavas, some metamorphic crystalline limestones and main constituent of uncompahgrite.

Physical Properties

Name: Mesolite (Zeolite Group)
Group: Zeolite
Formula: $Na_2Ca_2Al_6Si_9O_{30}+8H_2O$
Crystal System: Monoclinic
Color: Colorless, white
Opacity: Transparent
Luster: Vitreous, silky when fibrous
Streak: Colorless
SGLow: 2.259
SGHigh: 2.259
HardnessLow: 5
HardnessHigh: 5
Cleavage: 2
Direction: {101} and {10-1} perfect
Habit: Crystals acicular or fibrous, in divergent sprays, tufts or masses with radiating structure
Fracture: Uneven; brittle
Other: In cavities and veins within lavas, particularly basalts
Comments:

Optical Properties

Name: Mesolite (Zeolite Group)
Group: Zeolite
Formula: $Na_2Ca_2Al_6Si_9O_{30}+8H_2O$
Crystal System: Monoclinic
Color: Colorless
Form: Fibrous aggregates
Relief: Moderate, n<balsam
Birefringence: 2.259
2V: 80"
Nalpha or Nord.: 1.505
NBeta or Nextr.: 1.505
NGamma: 1.506
Optical Sign: Biaxial ppositive
Orientation: Partly length fast and partly length slow. r>v, v.strong
Pleochroism:
Twinning: Simple twins along {100}, not obvious
Cleavage: Two perfect along {110} and {11_0}
Extinction: Oblique, almost parallel 2"-5" in longitudinal sections
Alteration:
Features: Biaxial positive with large axial angle, small extinction angle
Occurrence: In cavities and veins within lavas, particularly basalts

Physical Properties

Name: Microcline

Group: Feldspar

Formula: KAlSi3O8

Crystal System: Triclinic

Color: White, grey, yellowish, reddish, green

Opacity: Transparent to translucent

Luster: Vitreous, often pearly on cleavage

Streak: White

SGLow: 2.55

SGHigh: 2.63

HardnessLow: 6

HardnessHigh: 6.5

Cleavage: 6

Direction: {001} and {010} perfect, {100}, {110}, {-110} and {-201} parting

Habit: Crystals short prismatic, blocky; or tabular; massive, cleavable to granular; twinning on Carlsbad, Manebach, Baveno laws

Fracture: Uneven; brittle

Other: Most abundant in felsic igneous rocks. Widespread occurrence

Comments: Dimorphous with orthoclase

Optical Properties

Name: Microcline

Group: Feldspar

Formula: KAlSi3O8

Crystal System: Triclinic

Color: Colorless

Form: Subhedral to anhedral coarse to fine crystals. Intergrown with microcline to form perthite

Relief: Low, n<balsam

Birefringence: Weak, 0.007

2V: 77"-84"

Nalpha or Nord.: 1.518-1.522

NBeta or Nextr.: 1.522-1.526

NGamma: 1.525-1.530

Optical Sign: Biaxial negative

Orientation: Length fast, Dispersion r>v

Pleochroism:

Twinning: Polysynthetic twinning almost always occurs, according to albite law or pericline law, giving quadrille structure with twinning at right angles.

Cleavage: Three directions, perfect {001}, distinct {010}, imperfect {110} and {11_0}

Extinction: Oblique on {001} >15", on {010} >5"

Alteration: Alters to clay

Features: Polysynthetic twinning distinctive. Extinction angle of 15" greater than albite

Occurrence: Most abundant in felsic igneous rocks. Widespread occurrence

Physical Properties

Name: Monazite
Group: Monazite
Formula: (Ce,La,Nd,Th)PO4
Crystal System: Monoclinic
Color: Reddish brown, brown, yellowish brown, pink, yellow, greenish, greyish white
Opacity: Transparent to subtranslucent
Luster: Resinous, waxy, or vitreous to subadamantine
Streak: White, slightly colored
SGLow: 4.6
SGHigh: 5.4
HardnessLow: 5
HardnessHigh: 5.5
Cleavage: 5
Direction: {100} distinct, {010} less distinct, {110}, {101} and {011} indistinct
Habit: Crystals thin to thick tabular, small, elongated on c-axis; also equant, wedge-shaped or prismatic; massive granular
Fracture: Conchoidal to uneven; brittle
Other: Occurs within pegmatites and granites and as a detrital mineral in sedimentary beds
Comments:

Optical Properties

Name: Monazite
Group: Monazite
Formula: (Ce,La,Nd,Th)PO4
Crystal System: Monoclinic
Color: Colorlesss to nuetral
Form: Small, euhedral crystals
Relief: V.high, n>balsam
Birefringence: Strong, 0.049-0.051. In cross sections weak, 0.001
2V: 6"-19"
Nalpha or Nord.: 1.786-1.800
NBeta or Nextr.: 1.788-1.801
NGamma: 1.837-1.849
Optical Sign: Biaxial positive
Orientation: Length slow
Pleochroism:
Twinning:
Cleavage: Five {100} distinct, {010} less distinct, {110}, {101} and {011} indistinct. Parting parallel to {001} often occurs
Extinction: Almost parallel 2" along length. Incomplete in sections parallel {001}
Alteration:
Features: Radioactive, euhedral crystals, birefringence
Occurrence: Occurs within pegmatites and granites and as a detrital mineral in sedimentary beds

Physical Properties

Name: Monticellite
Group: Olivine
Formula: CaMgSiO4
Crystal System: Orthorhombic
Color: Colorless, grey, greenish
Opacity: Transparent to translucent
Luster: Vitreous
Streak:
SGLow: 3.08
SGHigh: 3.27
HardnessLow: 5.5
HardnessHigh: 5.5
Cleavage: 1
Direction: {010} indistinct
Habit: Crystals prismatic; massive, granular; disseminated grains
Fracture: Brittle
Other: Occurs within contact metamorphic limestones and dolomites
Comments: Series with kirschsteinite

Optical Properties

Name: Monticellite
Group: Olivine
Formula: CaMgSiO4
Crystal System: Orthorhombic
Color: Colorless
Form: Anhedral to subhedral granular aggregates and euhedral prismatic crytals
Relief: High, n>balsam
Birefringence: Moderate, 0.014-0.018
2V: 75"-80"
Nalpha or Nord.: 1.641-1.651
NBeta or Nextr.: 1.646-1.662
NGamma: 1.655-1.669
Optical Sign: Biaxial negative
Orientation: Length slow, r>v
Pleochroism:
Twinning:
Cleavage: One imperfect {010}
Extinction: Parallel to outline and cleavage
Alteration: May be replaced by idocrase (vesuvianite)
Features: Similar to olivine and forsterite, but with lower birefringence and negative optical sign.
Occurrence: Occurs within contact metamorphic limestones and dolomites. Sometimes in nepheline basalt, alnoite and plozenite as overgrowths and rims in olivine.

Physical Properties

Name: Montmorillonite
Group: Smectite, Clay
Formula: (Na,Ca)0.3(Al,Mg)2Si4O10(OH)2+nH2O
Crystal System: Monoclinic
Color: White, grey, yellowish, greenish, pink
Opacity: Opaque
Luster: Dull
Streak:
SGLow: 2
SGHigh: 3
HardnessLow: 1
HardnessHigh: 2
Cleavage: 1
Direction: {001} perfect
Habit: Massive, fine-grained; clay-like
Fracture:
Other: Usually found as a diagenetic clay mineral in sedimentary beds. Alteration product oof volcanic ash and tuff
Comments: Synonomous with smectite

Optical Properties

Name: Montmorillonite
Group: Smectite, Clay
Formula: (Na,Ca)0.3(Al,Mg)2Si4O10(OH)2+nH2O
Crystal System: Monoclinic
Color: Pale pink, greenish or colorless
Form: Microcrystalline aggregates in the shape of shards or scale like crystals
Relief: Low, n>balsam
Birefringence: Moderate, 0.021
2V: 10"-25"
Nalpha or Nord.: 1.492
NBeta or Nextr.: 1.513
NGamma: 1.513
Optical Sign: Biaxial negative
Orientation:
Pleochroism:
Twinning:
Cleavage: One {001} perfect
Extinction:
Alteration:
Features: Microcrystalline aggregates of shards distinctive. XRD may be necessary.
Occurrence: Usually found as a diagenetic clay mineral in sedimentary beds. Alteration product oof volcanic ash and tuff

Physical Properties

Name: Mullite

Group: Silliminite

Formula: Al6Si2O13

Crystal System: Orthorhombic

Color: Colorless to pale pink

Opacity: Transparent to translucent

Luster: Vitreous

Streak:

SGLow: 3.03

SGHigh: 3.16

HardnessLow: 6

HardnessHigh: 7

Cleavage: 1

Direction: {010} distinct

Habit: Crystals prismatic

Fracture:

Other: Only occurs within xenoliths in igneous intrusions

Comments:

Optical Properties

Name: Mullite

Group: Sillimanite

Formula: Al6Si2O13

Crystal System: Orthorhombic

Color: Colorless

Form: Long prismatic crystals with square cross sections

Relief: High, n>balsam

Birefringence: Weak, 0.012

2V: 20"

Nalpha or Nord.: 1.642

NBeta or Nextr.: 1.644

NGamma: 1.654

Optical Sign: Biaxial positive

Orientation: Length slow, r>v

Pleochroism:

Twinning:

Cleavage: One distinct {010}

Extinction: Parallel in longitudinal sections and symmetrical in cross sections

Alteration:

Features: Almost identical to sillimanite but RI is lower. Occurrence distinctive.

Occurrence: Rare, only occurs within xenoliths of argillacious sediments in igneous intrusions

Physical Properties

Name: Muscovite

Group: Mica

Formula: KAl2(Si3Al)O10(OH,F)2

Crystal System: Monoclinic ps. hex.

Color: Colorless, grey, green, yellow, brown, violet, rose-red, deep ruby-red, pinkish red

Opacity: Transparent to translucent

Luster: Vitreous to pearly or silky

Streak: Colorless

SGLow: 2.77

SGHigh: 2.88

HardnessLow: 2.5

HardnessHigh: 4

Cleavage: 1

Direction: {001} perfect

Habit: Crystals tabular, hexagonal or diamond-shaped cross section; massive, scaly or lamellar; plumose or stellate aggregates

Fracture: Thin laminae flexible and elastic

Other: Widespread within felsic igneous rocks, gneisses, schists and low grade metamorphic rocks

Comments:

Optical Properties

Name: Muscovite

Group: Mica

Formula: KAl2(Si3Al)O10(OH,F)2

Crystal System: Monoclinic ps. hex.

Color: Colorless to pale green

Form: Thin, tabular crystals and scaly aggregates. Microcrystalline variety named sericite

Relief: Slight, n>balsam

Birefringence: Strong, 0.037-0.041

2V: 30"-40"

Nalpha or Nord.: 1.556-1.570

NBeta or Nextr.: 1.587-1.607

NGamma: 1.593-1.611

Optical Sign: Biaxial negative

Orientation: Length slow

Pleochroism: Some varieties show pleochroism

Twinning: According to mica law {110}

Cleavage: One perfect {001}

Extinction: Parallel up to 3"

Alteration:

Features: Cleavage, tabular cystals and strong birefringence distinctive. Similar to talc, but axial angle larger, physical examination should differentiate.

Occurrence: Widespread within felsic igneous rocks, gneisses, schists and low grade metamorphic rocks

Physical Properties

Name: Natrolite (Zeolite)
Group: Zeolite
Formula: Na2Al2Si3O10+2H2O
Crystal System: Orthorhombic
Color: Colorless, white, grey, yellowish, reddish
Opacity: Transparent to translucent
Luster: Vitreous to somewhat pearly
Streak: Colorless
SGLow: 2.20
SGHigh: 2.26
HardnessLow: 5
HardnessHigh: 5.5
Cleavage: 2
Direction: {110} perfect, {010} parting
Habit: Crystals prismatic, short or slender to acicular, striated vertically; massive, granular or compact; fibrous, radiating
Fracture: Uneven; brittle
Other: In cavities and veins within lavas, particularly basalts
Comments: Dimorphous with tetranatrolite

Optical Properties

Name: Natrolite (Zeolite)
Group: Zeolite
Formula: Na2Al2Si3O10+2H2O
Crystal System: Orthorhombic
Color: Colorless
Form: Long prismatic crystals or radiating fibrous aggregates. Cross sections almost square
Relief: Moderate, n<balsam
Birefringence: Weak, 0.012-0.013
2V: 60"-63"
Nalpha or Nord.: 1.473-1.480
NBeta or Nextr.: 1.476-1.482
NGamma: 1.485-1.493
Optical Sign: Biaxial positive
Orientation: Length slow
Pleochroism:
Twinning:
Cleavage: One parallel to length {110}
Extinction: Parallel to length, symmetrical in cross sections
Alteration:
Features: Radiating crystal aggregates, cleavge and extinction.
Occurrence: In cavities and veins within lavas, particularly basalts

Physical Properties

Name: Nepheline
Group: Feldspathoid
Formula: (Na,K)AlSiO4
Crystal System: Hexagonal
Color: Colorless, white, grey, yellowish, greenish, bluish grey, dark green, brownish red
Opacity: Transparent to nearly opaque
Luster: Vitreous to greasy
Streak: White
SGLow: 2.55
SGHigh: 2.67
HardnessLow: 5.5
HardnessHigh: 6
Cleavage: 2
Direction: {10-10} and {000-1} indistinct
Habit: Crystals simple hexagonal prisms; massive, compact; embedded grains
Fracture: Subconchoidal; brittle
Other: Characteristic of alkali igneous rocks
Comments:

Optical Properties

Name: Nepheline
Group: Feldspathoid
Formula: (Na,K)AlSiO4
Crystal System: Hexagonal
Color: Colorless to cloudy
Form: Euhedral hexagonal and prismatic phenocrysts and anhedral crytals if even grained rocks.
Relief: V. low, n=balsam or slightly higher
Birefringence: Weak, 0.003-0.004
2V:
Nalpha or Nord.: 1.527-1.543
NBeta or Nextr.: 1.530-1.547
NGamma:
Optical Sign: Uniaxial negative
Orientation:
Pleochroism:
Twinning:
Cleavage: One imperfect {101_0}
Extinction: Parallel, basal sections remain dark
Alteration: Alters easily to zeolites, sodalite, muscovite, cancrinite.
Features: Hexagonal and prismatic crystals distinctive
Occurrence: Characteristic of alkali igneous rocks

Physical Properties

Name: Nephrite
Group: Amphibole
Formula: $Ca_2(Mg,Fe)_5Si_8O_{22}(OH)_2$
Crystal System: Monoclinic
Color: Colorless, white, grey, pale greenish, pink, brown
Opacity: Transparent to translucent
Luster: Vitreous
Streak: Colorless
SGLow: 2.9
SGHigh: 3.2
HardnessLow: 5
HardnessHigh: 6
Cleavage: 2
Direction: {110} good, {100} parting
Habit: Crystals long-bladed or short and stout; fibrous or thin-columnar aggregates; massive, fibrous or granular
Fracture: Subconchoidal to uneven; brittle, some compact varieties tough
Other: Metamorphic mineral associated with serpentinites
Comments: Gem variety of compact actinolite or tremolite

Optical Properties

Name: Nephrite
Group: Amphibole
Formula: $Ca_2(Mg,Fe)_5Si_8O_{22}(OH)_2$
Crystal System: Monoclinic
Color: Colorless to grey
Form: Fibrous and imperfect prismatic crystals
Relief: High, n>balsam
Birefringence: Moderate, 0.022-0.027
2V: 79"-85"
Nalpha or Nord.: 1.600-1.628
NBeta or Nextr.: 1.613-1.644
NGamma: 1.633-1.701
Optical Sign: Biaxial negative
Orientation: Length slow
Pleochroism:
Twinning: Twins along {100} rare
Cleavage: Two 110} good, {100} parting. As for tremolite but rarely distinct due to felted fibres
Extinction: Parallel to 10"-20", often wavy and indistinct
Alteration: Alters to talc
Features: Smaller extinction angle and RI distinguishes from jadeite. Similar to other varieties of actinolite/tremolite, but often in felted masses.
Occurrence: Metamorphic mineral associated with serpentinites

Physical Properties

Name: Nosean (Sodalite Group)
Group: Sodalite, Feldspathoid
Formula: Na8Al6Si6O24(SO4)
Crystal System: Cubic
Color: Colorless, white, grey, bluish, brown, reddish, black
Opacity: Transparent to translucent
Luster: Vitreous
Streak: White
SGLow: 2.30
SGHigh: 2.40
HardnessLow: 5.5
HardnessHigh: 5.5
Cleavage: 1
Direction: {110} indistinct
Habit: Crystals dodecahedral; massive, granular
Fracture:
Other: Occurs in alkali igneous rocks
Comments:

Optical Properties

Name: Nosean (Sodalite Group)
Group: Sodalite, Feldspathoid
Formula: Na8Al6Si6O24(SO4)
Crystal System: Isometric cubic
Color: Colorless
Form: Crystals dodecahedral; massive, granular
Relief:
Birefringence:
2V:
Nalpha or Nord.:
NBeta or Nextr.:
NGamma:
Optical Sign:
Orientation:
Pleochroism:
Twinning:
Cleavage: One {110} indistinct
Extinction:
Alteration:
Features:
Occurrence: Occurs in alkali igneous rocks

Physical Properties

Name: Oligoclase

Group: Feldspar

Formula: (Na,Ca)(Si,Al)Si4O8

Crystal System: Triclinic

Color: Colorless, white, grey, greenish, yellowish, brown, reddish

Opacity: Transparent to translucent

Luster: Vitreous

Streak: White

SGLow: 2.63

SGHigh: 2.67

HardnessLow: 6

HardnessHigh: 6.5

Cleavage: 3

Direction: {001} perfect, {010} nearly perfect, {110} imperfect

Habit: Crystals tabular; commonly twinned on Carlsbad, albite or pericline laws; massive, cleavable, granular or compact

Fracture: Conchoidal to uneven; brittle

Other: Most abundant in felsic igneous rocks. Widespread occurrence

Comments:

Optical Properties

Name: Oligoclase

Group: Feldspar

Formula: (Na,Ca)(Si,Al)Si4O8 An10-30%

Crystal System: Triclinic

Color: Colorless

Form: Anhedral, subhedral and euhedral fine to coarse crystals

Relief: Low, n slightly <balsam, n=balsam, n>balsam

Birefringence: Weak, 0.0088-0.009

2V: 82"-90"

Nalpha or Nord.: 1.533-1.543

NBeta or Nextr.: 1.537-1.548t, {110} imperfe

NGamma: 1.542-1.551

Optical Sign: Biaxial positive or negative

Orientation: r>v weak

Pleochroism:

Twinning: According to albite law. Twins according to carlsbad and pericline law, as in albite

Cleavage: Four perfect {001}, distinct {010}, imperfect {110} and {11_0}

Extinction: Oblique to parallel 0"-12" in albite twins. 0"-3" in cleavage flakes {001}, 0"-15" {010}

Alteration: Alters to clay

Features: Extinction angles are distinctive. RI>balsam

Occurrence: Most abundant in felsic igneous rocks. Widespread occurrence

Physical Properties

Name: Olivine
Group: Olivine
Formula: (MgFe)2SiO4
Crystal System: Orthorhombic
Color: Green, lemon-yellow, white
Opacity: Transparent to translucent
Luster: Vitreous to greasy
Streak: Colorless
SGLow: 3.2
SGHigh: 4.3
HardnessLow: 6.5
HardnessHigh: 7
Cleavage: 2
Direction: {010} and {100} indistinct
Habit: Crystals thick tabular, vertically striated; massive, compact or granular; embedded grains
Fracture: Conchoidal; brittle
Other: Characteristic of mafic and ultramafic igneous and metamorphic rocks
Comments: Group name for forsterite-fayalite series

Optical Properties

Name: Olivine
Group: Olivine
Formula: (Mg,Fe)2SiO4
Crystal System: Orthorhombic
Color: Colorless
Form: Anhedral crystals with polygonal outlines
Relief: High, n>balsam
Birefringence: Strong, 0.037-0.041
2V: 70"-90"
Nalpha or Nord.: 1.651-1.681
NBeta or Nextr.: 1.670-1.706
NGamma: 1.689-1.718
Optical Sign: Biaxial positive and Biaxial negative
Orientation: Length slow, r<v
Pleochroism:
Twinning: Rare
Cleavage: Two {010} and {100} indistinct, irregular fractures
Extinction: Parallel to outline and cleavage
Alteration: Alters to antigorite (serpentine) and magnetite, obvious along fractures. In basaltic rocks alters to iron rich rims and brownish red iddingsite
Features: Polygonal outline, fractures, birefringence and extinction distinctive
Occurrence: Widespread mafic to ultrammafic igneous rocks, includimg basalts, gabbros and peridotites. Dunite consists entirely of olivine. A relict mineral in serpentinites.

Physical Properties

Name: Opal
Group: Quartz
Formula: SiO2.nH2O
Crystal System: Amorphous
Color: White, yellow, red, often with rich play of colors (opalescence)
Opacity: Transparent to translucent
Luster: Pearly, vitreous
Streak: White
SGLow: 2.0
SGHigh: 2.2
HardnessLow: 5.5
HardnessHigh: 6.5
Cleavage: None
Direction:
Habit: None, usually massive. Also botryoidal, reniform, stalactitic
Fracture: Conchoidal
Other: Low temperature mineral that fills cavities and fractures in rocks. Replaces cells in wood and shells in fossils. Forms amygdales in basalts and rhyolites.
Comments: Coober Pedy, Australia is main source of opals. Formed in a desert environment by percolating hydrothermal solutions near the surface.

Optical Properties

Name: Opal
Group: Opal
Formula: SiO2(H2O)x
Crystal System: Amorphous to microcrystalline
Color: Colorless to pale grey, brown. Gem opal displays multi color array
Form: Colloform crusts, veins and cavity fillings
Relief: High, n< balsam
Birefringence: None
2V:
Nalpha or Nord.: 1.40-1.46
NBeta or Nextr.:
NGamma:
Optical Sign:
Orientation:
Pleochroism:
Twinning:
Cleavage: None
Extinction:
Alteration:
Features: Play of colors
Occurrence: Secondary mineral in cavities and seams.

Physical Properties

Name: Orthoclase
Group: Feldspar
Formula: KAlSi3O8
Crystal System: Monoclinic
Color: Colorless, white, grey, yellow, reddish, greenish
Opacity: Transparent to translucent
Luster: Vitreous to pearly
Streak: White
SGLow: 2.55
SGHigh: 2.63
HardnessLow: 6
HardnessHigh: 6.5
Cleavage: 6
Direction: {001} and {010} perfect, {100}, {110}, {-110} and {-201} partings
Habit: Crystals short prismatic, blocky, may have orthorhombic or tetragonal aspect; massive, cleavable to granular; twinned
Fracture: Conchoidal to uneven; brittle
Other: Widespread within felsic igneous rocks, metamorphic gneisses and sedimentary sandstones
Comments: Dimorphous with microcline; series with celsian and hyalophane

Optical Properties

Name: Orthoclase
Group: Feldspar
Formula: KAlSi3O8
Crystal System: Monoclinic
Color: Colorless
Form: Subhedral and anhedral coarse to fine crystals, phenocrysts and in spherulites
Relief: Low, n<balsam
Birefringence: Weak, 0.008
2V: 69"-72"
Nalpha or Nord.: 1.518
NBeta or Nextr.: 1.524
NGamma: 1.526
Optical Sign: Biaxial negative
Orientation: Legth fast, r>v
Pleochroism:
Twinning: Simple twinning according to carlsbad law
Cleavage: Three directions, perfect {001}, distinct {010}, imperfect {110}
Extinction: Parallel on {001}, oblique 5"-12" on {010}, increasing with soda content
Alteration: Altered to kaolinite. Dimorphic with sanidine
Features: Distinguished from sanidine by large axial angle. Twinning distinctive.
Occurrence: Widespread within felsic igneous rocks, metamorphic gneisses and sedimentary sandstones

Optical Properties

Name: Palagonite

Group: Mineraloid

Formula: $SiO_2Al_2O_3Fe_2O_3FeO,MgOCaO,H_2O$

Crystal System: Amorphous

Color: Yellow to yellowish brown. Also brown or green

Form: Rims and zones around glass fragments or massive. oolitic structure due to filling of micro-vesicles

Relief: Low to medium, n< or n> balsam

Birefringence: None to v.weak

2V:

Nalpha or Nord.: 1.47-1.63

NBeta or Nextr.:

NGamma:

Optical Sign:

Orientation:

Pleochroism:

Twinning:

Cleavage:

Extinction:

Alteration:

Features: Similar to opal, collophane and volcanic glass. Intermediate RI between collophane and glass.

Occurrence: Occurs in volcanic igneous rocks, such as tuffs and basaltic breccias, formed by hydration of fragmented basaltic glass.

Physical Properties

Name: Palygorskite
Group: Clay
Formula: (Mg,Al)2Si4O10(OH)+4H2O
Crystal System: Monoclinic and orthorhombic
Color: White, grey
Opacity: Translucent
Luster: Dull
Streak:
SGLow: 2.217
SGHigh: 2.217
HardnessLow: 1
HardnessHigh: 1
Cleavage: 1
Direction: {110} easy
Habit: Crystals lath-like, elongated, in bundles; forms flexible sheets of interlaced fibers, resembles leather or parchment
Fracture: Tough
Other: Chief component of Fullers Earth
Comments: H=soft

Optical Properties

Name: Palygorskite
Group: Clay
Formula: (Mg,Al)2Si4O10(OH)+4H2O
Crystal System: Monoclinic and orthorhombic
Color: Colorless to pale yellow or green
Form: Compact fine aggregates
Relief: Low, n<balsam
Birefringence: Moderate, 0.023
2V:
Nalpha or Nord.: 1.510
NBeta or Nextr.:
NGamma: 1.533
Optical Sign:
Orientation:
Pleochroism:
Twinning:
Cleavage: One {110} perfect
Extinction: Aggregates show uniform mass extinction
Alteration:
Features: Similar to montmorillinite, may require XRD.
Occurrence: Chief component of Fullers Earth

Physical Properties

Name: Pectolite

Group: Pyroxene

Formula: NaCa2Si3O8(OH)

Crystal System: Triclinic

Color: Colorless, white

Opacity: Transparent to translucent

Luster: Vitreous or silky

Streak: White

SGLow: 2.74

SGHigh: 2.88

HardnessLow: 4.5

HardnessHigh: 5

Cleavage: 2

Direction: {001} and {100} perfect

Habit: Crystals tabular or acicular, elongated along b-axis; in radiating aggregates, globular

Fracture: Splintery, uneven

Other: Chiefly occurs in cavities of volcqnic rocks with zeolites

Comments: Series with serandite

Optical Properties

Name: Pectolite

Group: Pyroxene

Formula: NaCa2Si3O8(OH)

Crystal System: Triclinic

Color: Colorless

Form: Crystals tabular or acicular, elongated along b-axis; in radiating aggregates, globular

Relief:

Birefringence:

2V:

Nalpha or Nord.:

NBeta or Nextr.:

NGamma:

Optical Sign:

Orientation:

Pleochroism:

Twinning:

Cleavage: Two {001} and {100} perfect

Extinction: Parallel or almost parallel in longitudinal sections, oblique in cross sections

Alteration:

Features: Series with serandite

Occurrence: Chiefly occurs in cavities of volcanic rocks with zeolites

Optical Properties

Name: Penninite

Group: Chlorite

Formula: $Mg_5(Al,Fe)(OH)_8(Al,Si)_4)_{10}$

Crystal System: Monoclinic

Color: Greenish

Form: Thick tabular pseudo-hexagonal sections

Relief: Fair, n>balsam

Birefringence: V. weak, 0.001-0.004

2V: 0"-20"

Nalpha or Nord.: 1.575-1.582

NBeta or Nextr.: 1.576-1.582

NGamma: 1.576-1.583

Optical Sign: Biaxial ppositive or negative

Orientation: Length fast or length slow

Pleochroism: Pleochroic from green - nearly colorless and green - brownish red, Anomolous "berlin blue" common

Twinning: Polysynthetic according to pennine law {001}

Cleavage: One perfect {001}

Extinction: Parallel or almost so

Alteration:

Features: Similar to other chlorites, but parallel extinction and anomolous birefringence are distinctive

Occurrence: Alteration product of garnets etc. Usually in seams and cavities.

Optical Properties

Name: Periclase

Group: Oxides

Formula: MgO

Crystal System: Isometric, cubic

Color: Colorless

Form: Small crystals or anhedral crystal aggregates

Relief: High, n>balsam

Birefringence: Nil, dark between crossed polars

2V:

Nalpha or Nord.: 1.738 to 1.760

NBeta or Nextr.:

NGamma:

Optical Sign:

Orientation:

Pleochroism:

Twinning:

Cleavage:

Extinction:

Alteration: Usually altered to brucite

Features: Cubic cleavage, high relief and isotropism

Occurrence: Usually in metamorphic limestones, quite rare. A metamorphic mineral.

Physical Properties

Name: Perovskite
Group: Spinel
Formula: CaTiO3
Crystal System: Orthorhombic ps. cub.
Color: Black, dark brown, amber to yellow
Opacity: Transparent to opaque
Luster: Adamantine, dull, metallic
Streak: Colorless to pale grey
SGLow: 4.01
SGHigh: 4.01
HardnessLow: 5.5
HardnessHigh: 5.5
Cleavage: 1
Direction: {001} imperfect
Habit: Crystals pseudocubic, often highly modified, or pseudo-octahedral; massive granular or reniform
Fracture: Subconchoidal to uneven; brittle
Other: In basic igneous rocks, chlorite and talc schists
Comments:

Optical Properties

Name: Perovskite
Group: Spinel
Formula: CaTiO3
Crystal System: Orthorhombic ps. cub.
Color: Yellow to brown
Form: Minute cubic crystals
Relief: V.high, n>balsam,adamantine luster in reflected light
Birefringence: Nil to .0002, larger crystals show weak birefringence
2V:
Nalpha or Nord.:
NBeta or Nextr.:
NGamma:
Optical Sign:
Orientation:
Pleochroism:
Twinning:
Cleavage: One {001} imperfect
Extinction:
Alteration:
Features: Color, crystals, cleavage
Occurrence: In basic igneous rocks, chlorite and talc schists

Physical Properties

Name: Phlogopite
Group: Mica
Formula: KMg3Si3AlO10(F,OH)2
Crystal System: Monoclinic
Color: Yellowish brown to brownish red, colorless, white, greenish
Opacity: Transparent to translucent
Luster: Pearly, submetallic on cleavage
Streak: Colorless
SGLow: 2.76
SGHigh: 2.90
HardnessLow: 2
HardnessHigh: 2.5
Cleavage: 1
Direction: {001} perfect
Habit: Crystals prismatic, tapered; plates; scales
Fracture: Thin laminae flexible and elastic, tough
Other: Characteristic of metamorphic limestones
Comments: Series with biotite

Optical Properties

Name: Phlogopite
Group: Mica
Formula: KMg3Si3AlO10(F,OH)2
Crystal System: Monoclinic
Color: Colorless to pale brown
Form: Short prismatic and thick six sided tabular crystals
Relief: Fair, n>balsam
Birefringence: Strong, 0.044-0.047
2V: 0"-10"
Nalpha or Nord.: 1.551-1.562
NBeta or Nextr.: 1.598-1.606
NGamma: 1.598-1.606
Optical Sign: Biaxial negative
Orientation: Length slow, r>v weak
Pleochroism: Weak pleochroism
Twinning: According to mica law
Cleavage: One perfect {001}
Extinction: Parallel, upto 5"
Alteration:
Features: Similar to biotite but lighter color and weaker adsorption. Similar to muscovite but smaller axial angle. occurrence may be distinctive.
Occurrence: Characteristic of metamorphic limestones

Descriptive Handbook of Rock Forming Minerals

Physical Properties

Name: Piedmontite
Group: Epidote
Formula: Ca2(Al,Mn,Fe)3(SiO4)3(OH)
Crystal System: Monoclinic
Color: Reddish brown to reddish black, purplish-red, straw-yellow
Opacity: Translucent to nearly opaque
Luster:
Streak:
SGLow: 3.45
SGHigh: 3.52
HardnessLow: 6
HardnessHigh: 6
Cleavage: 1
Direction: {001} perfect
Habit: Crystals prismatic or acicular; massive, coarse to fine granular; fibrous
Fracture: Uneven; brittle
Other: Mostly within metamorphic schists and gneisses
Comments:

Optical Properties

Name: Piedmontite
Group: Epidote
Formula: Ca2(Al,Mn,Fe)3(SiO4)3(OH)
Crystal System: Monoclinic
Color: Yellow, orange, red, violet
Form: Short prismatic crystals, columnar aggregates and six sided cross sections, as in epidote
Relief: High, n>balsam
Birefringence: Very strong, 0.061-0.082
2V: 56"-86"
Nalpha or Nord.: 1.745-1.758
NBeta or Nextr.: 1.764-1.789
NGamma: 1.806-1.832
Optical Sign: Biaxial positive
Orientation: Difficult, r>v strong
Pleochroism: Strong pleochroism from yellow to orange - amethyst to violet - carmine to deep red
Twinning:
Cleavage: One perfect parallel to {001}
Extinction: Parallel
Alteration:
Features: Color and pleochroism distinctive
Occurrence: Mostly within metamorphic schists and gneisses. Manganian variety of eppidote.

Page 135

Optical Properties

Name: Pigeonite

Group: Pyroxene

Formula: CaMg(SiO3)2.(Mg,Fe)SiO3

Crystal System: Monoclinic

Color: Colorless to neutral

Form: Usually as anhedral crystals

Relief: High, n>balsam

Birefringence: Moderate, 0.021-0.033

2V: 0"-40"

Nalpha or Nord.: 1.680-1.718

NBeta or Nextr.: 1.698-1.725

NGamma: 1.719-1.744

Optical Sign: Biaxial positive

Orientation: Length slow

Pleochroism: May show faint pleochroism

Twinning: Polysynthitic twinning along {100} characteristic

Cleavage: Two directions at 87" and 93"

Extinction: Oblique 22"-45", increasing with clino-enstite content

Alteration:

Features: Twinning distinctive

Occurrence: Widespread in mafic to ultramafic volcanic rocks, including basalts and pyroxenites.Usually as groundmass, rarel as phenocrysts. Also in dolerites and diabases

Optical Properties

Name: Polyhalite

Group: Evaporite

Formula: K2MgCa2(SO4)4.2H2O

Crystal System: Triclinic

Color: Colorless to reddish(hematite)

Form: Granular or fibrous

Relief: Low, n>balsam

Birefringence: Moderate, 0.019

2V: 70"

Nalpha or Nord.: 1.548

NBeta or Nextr.: 1.562

NGamma: 1.567

Optical Sign: Biaxial negative

Orientation:

Pleochroism:

Twinning: Plysynthetic with {010} common

Cleavage: Two parallel to {100}. Parting parallel to {010}

Extinction: Oblique

Alteration:

Features: Birefringence and refractive index are greater than gypsum. Soluble in water

Occurrence: Occurs with gypsum, anhydrite, halite in sedimentary evaporite deposits

Physical Properties

Name: Prehnite

Group: Serpentine

Formula: $Ca_2Al_2Si_3O_{10}(OH)_2$

Crystal System: Orthorhombic

Color: Pale to dark green, yellow, grey, white, colorless

Opacity: Transparent to translucent

Luster: Vitreous to somewhat pearly

Streak: Colorless

SGLow: 2.90

SGHigh: 2.95

HardnessLow: 6

HardnessHigh: 6.5

Cleavage: 2

Direction: {001} distinct, {110} indistinct

Habit: Crystals tabular, prismatic or steep pyramidal, in tabular groups or barrel-shaped aggregates; massive; botryoidal

Fracture: Uneven; brittle

Other: Found in cavities and seams of igneous rocks, sometimes in veins. Associated with calcite, quartz, datolite and

Comments:

Optical Properties

Name: Prehnite

Group: Serpentine

Formula: $Ca_2Al_2Si_3O_{10}(OH)_2$

Crystal System: Orthorhombic

Color: Colorless

Form: Sheaflike aggregates, almost spherulitic

Relief: High, n>balsam

Birefringence: Moderate to strong, 0.020-0.033. Anomolous in some varieties

2V: Variable

Nalpha or Nord.: 1.615-1.635

NBeta or Nextr.: 1.624-1.642

NGamma: 1.645-1.665

Optical Sign: Biaxial positive

Orientation: Length fast, r>v weak

Pleochroism:

Twinning: Polysynthetic in two directions at right angles

Cleavage: One good {001}

Extinction: Parallel to cleavage

Alteration:

Features: Similar to lawsonite, but higher birefringence and RI

Occurrence: Found in cavities and seams of igneous rocks, sometimes in veins. Associated with calcite, quartz, datolite and zeolites.

Physical Properties

Name: Pumpellyite-[Fe]
Group: Epidote
Formula: $Ca_2FeAl_2(SiO_4)(Si_2O_7)(OH)_2 + H_2O$
Crystal System: Monoclinic
Color: Blue-green, green, brown
Opacity: Translucent
Luster: Vitreous
Streak: White
SGLow: 3.18
SGHigh: 3.23
HardnessLow: 6
HardnessHigh: 6
Cleavage: 2
Direction: {001} and {100} distinct
Habit: Crystals fibrous or narrow plates; stellate clusters; dense mats
Fracture: Uneven, brittle
Other: A low grade metamorphic mineral found in pelitic meta-sediments
Comments: Series with pumpellyite-(Mg) and with julgoldite-Fe+2

Optical Properties

Name: Pumpellyite-[Fe]
Group: Epidote
Formula: $Ca_2FeAl_2(SiO_4)(Si_2O_7)(OH)_2 + H_2O$
Crystal System: Monoclinic
Color: Colorless, pale green, pale brown, yellow.
Form: Subhedral prismatic and fibrous crystals
Relief: high, n>balsam
Birefringence: Moderate, 0.014-0.020
2V: 26"-84"
Nalpha or Nord.: 1.674-1.702
NBeta or Nextr.: 1.675-1.715
NGamma: 1.688-1.722
Optical Sign: Biaxial positive
Orientation: r<v,
Pleochroism: Strong pleochroism from colorless or pale yellow-brown or pale greenish yellow - bluish green or pale green or brownish yellow - colorless or pale yellowish brown or brownish yellow
Twinning: Penetration twins infrequent
Cleavage: Two directions distinct {001} and {100}
Extinction: Oblique 22" on longitudinal sections
Alteration:
Features: Color, pleochroism, extinction angle of 22" distinctive
Occurrence: A low grade metamorphic mineral found in pelitic meta-sediments and low grade schists.

Physical Properties

Name: Pyrite

Commodity: Gold (Au)

Formula: FeS2

Crystal System: Cubic

Color: Brass-yellow

Opacity: Opaque

Luster: Metallic

Streak: Empty

SGLow: 5.018

SGHigh: 5.028

HardnessLow: 6

HardnessHigh: 6.5

Cleavage: 3

Direction: {100} and {311} indistinct, {110} occasional parting

Habit: Crystals cubic, pyritohedral, octahedral and combinations, abnormally developed, rarely acicular; massive; granular

Fracture: Conchoidal to uneven; brittle

Other: Monetary use, jewellery, electronics and for decoration

Comments: Associated with other sulphide minerals and gold in hypothermal to epithermal quartz veins. Sometimes contains small amounts of gold and is mined as an ore of gold

Optical Properties

Name: Pyrite

Group: Pyrite

Formula: FeS2

Crystal System: Isometric, cubic

Color: Brass yellow in reflected light

Form: Euhedral cubes with square, rectangular and triangular outlines. Also in grains , masses and veins

Relief:

Birefringence: Blue-green-red-brown in reflected light

2V:

Nalpha or Nord.:

NBeta or Nextr.:

NGamma:

Optical Sign:

Orientation:

Pleochroism:

Twinning:

Cleavage: None

Extinction:

Alteration: Alters to limonite

Features: Brass yellow and cubic crystals. Similar to chalcopyrite, which is darker yellow, and pyrrhotite, which is bronze.

Occurrence: Widespread, most common in veins and hydrothermal replacement deposits.

Physical Properties

Name: Pyromorphite
Group: Apatite
Formula: Pb5(PO4)3Cl
Crystal System: Hexagonal
Color: Green, yellow, orange, brown, grey, colorless, white
Opacity: Transparent to translucent
Luster: Subadamantine to resinous
Streak: White
SGLow: 7.04
SGHigh: 7.04
HardnessLow: 3.5
HardnessHigh: 4
Cleavage: 1
Direction: {10-11} trace
Habit: Crystals short prisms, hexagonal, often barrel-shaped, cavernous; equant, tabular or pyramidal; granular; globular
Fracture: Subconchoidal to uneven; brittle
Other: Occurs in hydrothermal replacement deposits
Comments:

Optical Properties

Name: Pyromorphite
Group: Apatite
Formula: Pb5(PO4)3Cl
Crystal System: Hexagonal
Color: White
Form: Crystals short prisms, hexagonal, often barrel-shaped, cavernous; equant, tabular or pyramidal; granular; globular
Relief:
Birefringence:
2V:
Nalpha or Nord.:
NBeta or Nextr.:
NGamma:
Optical Sign:
Orientation:
Pleochroism:
Twinning:
Cleavage:
Extinction:
Alteration:
Features:
Occurrence: Occurs in hydrothermal replacement deposits

Physical Properties

Name: Pyrope
Group: Garnet
Formula: Mg3Al2(SiO4)3
Crystal System: Cubic
Color: Pinkish red, purplish red, orange-red, crimson, nearly black
Opacity: Transparent to translucent
Luster: Vitreous
Streak: Colorless
SGLow: 3.5
SGHigh: 3.8
HardnessLow: 7
HardnessHigh: 7.5
Cleavage: 1
Direction: {110} parting sometimes distinct
Habit: Crystals dodecahedrons or trapezohedrons, rare; rounded pebbles; embedded grains
Fracture: Conchoidal; brittle
Other: Mainly occurs within peridotites and metamorphic serpentinites
Comments: Series with almandine and with knorringite

Optical Properties

Name: Pyrope
Group: Garnet
Formula: Mg3Al2(SiO4)3
Crystal System: Isometric, cubic
Color: Colorless
Form: Euhedral dodecahedrons in six sided trapezohedrons in eigth sided cross sections. Plygonal grains and aggregates.
Relief: V high, n>balsam
Birefringence: Isotropic, may show weak birefringence
2V:
Nalpha or Nord.: 1.741-1.760
NBeta or Nextr.:
NGamma:
Optical Sign:
Orientation:
Pleochroism:
Twinning:
Cleavage: May have parting parallel to {110}, irregular fractures
Extinction:
Alteration:
Features: Similar to spinel, which is octohedral.
Occurrence: Mainly occurs within peridotites and metamorphic serpentinites

Physical Properties

Name: Pyrophyllite
Group: Mica
Formula: Al2Si4O10(OH)2
Crystal System: Monoclinic and triclinic
Color: White, greyish white, yellowish, pale blue, greenish, greyish or brownish green
Opacity: Transparent to translucent
Luster: Pearly to dull and glistening
Streak: White
SGLow: 2.65
SGHigh: 2.90
HardnessLow: 1
HardnessHigh: 2
Cleavage: 1
Direction: {001} perfect
Habit: Crystals tabular, elongated, subhedral, curved and distorted; foliated; radiated lamellar or fibrous; granular to compact
Fracture: Laminae flexible, inelastic
Other: A hydrothermal alteration product occuring in mketamorphic rocks with andalusite, sillimanite and kyanite
Comments:

Optical Properties

Name: Pyrophyllite
Group: Brittle Mica
Formula: Al2Si4O10(OH)2
Crystal System: Monoclinic and triclinic
Color: Colorless
Form: Tabular, elongated crystals, fine aggregates and curved, distorted crystals, radial aggregates.
Relief: Low to moderate, n>balsam
Birefringence: Strong, 0.048, sections parallel to cleavage give first order grey colors
2V: 53"-60"
Nalpha or Nord.: 1.552
NBeta or Nextr.: 1.588
NGamma: 1.600
Optical Sign: Biaxial negative
Orientation: Length slow
Pleochroism:
Twinning: Plysynthetic according to mica law but not distinct
Cleavage: One perfect {001}
Extinction: Parallel to cleavage and outline
Alteration:
Features: Microcrystalline variety similar to sericite and talc, xrd may be necessary to distinguish.
Occurrence: A hydrothermal alteration product occuring in metamorphic rocks with andalusite, sillimanite and kyanite

Physical Properties

Name: Quartz
Group: Quartz
Formula: SiO2
Crystal System: Trigonal
Color: Colorless, white, grey, yellow to brown to black, violet, pink
Opacity: Transparent to translucent
Luster: Vitreous, sometimes greasy or waxy
Streak: White
SGLow: 2.655
SGHigh: 2.655
HardnessLow: 7
HardnessHigh: 7
Cleavage: 0
Direction: ({10-11}, {01-11}, {10-10}, {0001},{11-20}, {11-21} and {51-61} seldom distinct)
Habit: Crystals short to long prismatic, elongated along c-axis, hexagonal, horizontally striated, bent, distorted, skeletal
Fracture: Subconchoidal to conchoidal; brittle
Other: Most widespread of all minerals
Comments: The most widespread rock forming mineral

Optical Properties

Name: Quartz
Group: Quartz
Formula: SiO2
Crystal System: Hexagonal
Color: Colorless
Form: Euhedral hexagonal crystals, anhedral grains, veins, deisseminated grains. Anhedral fne to coarse crystals
Relief: V.low, n>balsam
Birefringence: Weak, 0.009
2V:
Nalpha or Nord.: 1.5442
NBeta or Nextr.: 1.5533
NGamma:
Optical Sign: Uniaxial positive
Orientation: Length slow in euhedral crystals
Pleochroism:
Twinning: Common, but rarely shows in thin section
Cleavage: None
Extinction: Parallel in euhedral crystals, wavy extinction in large crsytals due to strain.
Alteration:
Features: Absence of cleavage, hexagonal crystals, first order birefringence, uniaxial p.sitive
Occurrence: Most widespread of all minerals

Physical Properties

Name: Rhodonite
Group: Pyroxene
Formula: (Mn,Fe,Mg,Ca)SiO3
Crystal System: Triclinic
Color: Pink to rose-red to brownish red, yellow, grey
Opacity: Transparent to translucent
Luster: Vitreous, somewhat pearly on cleavage
Streak: Colorless
SGLow: 3.57
SGHigh: 3.76
HardnessLow: 5.5
HardnessHigh: 6.5
Cleavage: 3
Direction: {110} and {1-10} perfect; {001} good
Habit: Crystals tabular, usually rough, rounded edges; massive, cleavable to compact; granular, coarse to fine
Fracture: Conchoidal to uneven; tough when compact
Other: Occurs in metamorphic hornfels and hydrothermal replacement deposits
Comments:

Optical Properties

Name: Rhodonite
Group: Pyroxene
Formula: (Mn,Fe,Mg,Ca)SiO3
Crystal System: Triclinic
Color: Colorless
Form: Crystals tabular, usually rough, rounded edges; massive, cleavable to compact; granular, coarse to fine
Relief:
Birefringence:
2V:
Nalpha or Nord.:
NBeta or Nextr.:
NGamma:
Optical Sign:
Orientation:
Pleochroism:
Twinning:
Cleavage: Three {110} and {1-10} perfect; {001} good
Extinction: Parallel or almost parallel in longitudinal sections, oblique in cross sections
Alteration:
Features:
Occurrence: Occurs in metamorphic hornfels and hydrothermal replacement deposits

Physical Properties

Name: Riebeckite (Crocidolite)
Group: Amphibole
Formula: Na2(Fe,Mg)3Fe2Si8O22(OH)2
Crystal System: Monoclinic
Color: Dark blue to black
Opacity: Translucent to nearly opaque
Luster: Vitreous, silky
Streak: White to blue-grey
SGLow: 3.32
SGHigh: 3.382
HardnessLow: 5
HardnessHigh: 5
Cleavage: 1
Direction: {110} perfect
Habit: Crystals long prismatic, striated along elongation; massive, fibrous to asbestiform, columnar, granular
Fracture: Uneven; brittle
Other: Characteristic of alkali igneous rocks, including alkali granites, syenites and trachytes. Also in metamorphic banded iron formations.
Comments: Series with magnesioriebeckite. Main asbestiform mineral.

Optical Properties

Name: Riebeckite (Crocidolite)
Group: Amphibole
Formula: Na2(Fe,Mg)3Fe2Si8O22(OH)2
Crystal System: Monoclinic
Color: Dark blue
Form: Subhedral prismatic crystals and fibrous, asbestiform aggregates
Relief: High, n>balsam
Birefringence: Weak, 0.004
2V: Large
Nalpha or Nord.: 1.697
NBeta or Nextr.: 1.695
NGamma: 1.697
Optical Sign: Biaxial negative
Orientation: Length fast, r>v strong
Pleochroism: Strong pleochroism from deep b;lue-lighter blue-greenish
Twinning:
Cleavage: In two directions {110} at 56" and 124"
Extinction: Oblique 5" in longitudinal sections
Alteration: Alters to tiger-eye
Features: Color, pleochroism and extinction angle are distinctive
Occurrence: Characteristic of alkali igneous rocks, including alkali granites, syenites and trachytes. Also in metamorphic banded iron formations.

Physical Properties

Name: Rubellite (Tourmaline)
Group: Tourmaline
Formula: NaFe3Al6(BO3)3Si6O18(OH)4
Crystal System: Trigonal
Color: Black, brownish black, bluish black
Opacity: Translucent to opaque
Luster: Vitreous to resinous
Streak: Colorless
SGLow: 3.10
SGHigh: 3.25
HardnessLow: 7
HardnessHigh: 7
Cleavage: 2
Direction: {11-20} and {10-11} indistinct
Habit: Crystals short to long prismatic, vertically striated, or acicular to fibrous, thin tablets rare; radiating groups; massive
Fracture: Conchoidal to uneven; brittle
Other: Mainly found within ggranite pegmatites, granites and high temperature veins
Comments: Pink to red variety

Optical Properties

Name: Rubellite (Tourmaline)
Group: Tourmaline
Formula: NaFe3Al6(BO3)3Si6O18(OH)4
Crystal System: Trigonal
Color: Colorless
Form: Crystals short to long prismatic, vertically striated, or acicular to fibrous, thin tablets rare; radiating groups; massive
Relief:
Birefringence:
2V:
Nalpha or Nord.:
NBeta or Nextr.:
NGamma:
Optical Sign:
Orientation:
Pleochroism:
Twinning:
Cleavage: Two {11-20} and {10-11} indistinct
Extinction: Parallel, cross sections remain dark
Alteration:
Features: Pink to red variety
Occurrence: Mainly found within ggranite pegmatites, granites and high temperature veins

Physical Properties

Name: Rutile

Group: Rutile

Formula: TiO_2

Crystal System: Tetragonal

Color: Reddish brown to red, yellow, orange-yellow, bluish, greyish black to black, greenish

Opacity: Transparent to translucent

Luster: Adamantine, submetallic

Streak: Pale brown to yellowish

SGLow: 4.23

SGHigh: 4.23

HardnessLow: 6

HardnessHigh: 6.5

Cleavage: 5

Direction: {110} and {100} distinct, {111} traces, {092} and {011} parting

Habit: Crystals short prismatic, often striated along c-axis; slender prismatic to acicular; massive, compact to granular

Fracture: Conchoidal to uneven; brittle

Other: Widespread accessory mineral in metamorphic rocks. Common detrital mineral

Comments: Trimorphous with anatase and brookite

Optical Properties

Name: Rutile

Group: Rutile

Formula: TiO_2

Crystal System: Tetragonal

Color: Pale brown to yellowish

Form: Small prismatic to acicular crystals

Relief: V. high, n>balsam, adamantine luster in reflected light

Birefringence: Extreme, 0.286-0.287

2V: 4.23

Nalpha or Nord.: 2.603-2.616

NBeta or Nextr.: 2.889-2.903

NGamma:

Optical Sign: Uniaxial Positive

Orientation:

Pleochroism:

Twinning: Knee shaped twins are common, Capillary crystals also common, especially in quartz.

Cleavage: Five {110} and {100} distinct, {111} traces, {092} and {011} parting. Parallel to length of the crystal {110}

Extinction:

Alteration:

Features: The color, together with v.high relief, birefringence.

Occurrence: Widespread accessory mineral in metamorphic rocks. Common detrital mineral

Physical Properties

Name: Sanidine

Group: Feldspar

Formula: $(K,Na)(Si,Al)4O8$

Crystal System: Monoclinic

Color: Colorless, whitish, tan, pinkish

Opacity: Transparent to translucent

Luster: Vitreous

Streak: White

SGLow: 2.56

SGHigh: 2.62

HardnessLow: 6

HardnessHigh: 6

Cleavage: 6

Direction: {001} and {010} perfect; {100}, {110}, {-110} and {-210} partings

Habit: Crystals prismatic, often tabular

Fracture:

Other: Characteristic of felsic volcanic rocks. Widespread occurrence

Comments: K-Na feldspar with disordered Al-Si arrangement

Optical Properties

Name: Sanidine

Group: Feldspar

Formula: $(K,Na)(Si,Al)4O8$

Crystal System: Monoclinic

Color: Colorless

Form: Subhedral to eudral monoclinic prisms as phenocrysts

Relief: Low, n<balsam

Birefringence: Weak, 0.007

2V: 0"-12"

Nalpha or Nord.: 1.517-1.520

NBeta or Nextr.: 1.523-1.525

NGamma: 1.524-1.526

Optical Sign: Biaxial negative

Orientation: (1)axial plane {010} r>v, (2) axial plane perpendicular to {010} r<v

Pleochroism:

Twinning: Simple twinning according to carlsbad law

Cleavage: Two directions, perfect {001}, distinct {010}, may be parting parallel to {100}

Extinction: Parallel on {001}, >5" on {010}, Sections normal to optic axis are almost dark, due to small axial angle

Alteration:

Features: Small axial angle differentaites from orthoclase

Occurrence: Considered characteristic of felsic volcanic rocks. Widespread occurrence

Physical Properties

Name: Scapolite
Group: Scapolite
Formula: (Na,Ca,K)4[Al3(Al,Si)3Si6)24](Cl,CO3,SO4,OH)
Crystal System: Tetragonal
Color: White, bluish-grey, grenish-yellow, yellow, pink, violet, brown, orange-brown
Opacity: Translucent
Luster: Non metallic
Streak: White
SGLow: 2.5
SGHigh: 2.62
HardnessLow: 5
HardnessHigh: 6
Cleavage: 2
Direction: {100}, {110} good
Habit: Prismatic
Fracture: Conchoidal to uneven
Other: Occurs within contact metamorphic limestones with idocrase, diopside and garnet. Also in some gneisses
Comments: A group name

Optical Properties

Name: Scapolite
Group: Scapolite
Formula: (Na,Ca,K)4[Al3(Al,Si)3Si6)24](Cl,CO3,SO4,OH)
Crystal System: Tetragonal
Color: Colorless
Form: Mostly as columnar aggregates, usually large
Relief: Low to fair, n>balsam
Birefringence: Weak to strong, 0.010-0.036, increasing with calcium content
2V: 2.62
Nalpha or Nord.: 1.540-1.571
NBeta or Nextr.: 1.550-1.607
NGamma:
Optical Sign: Uniaxial negative
Orientation:
Pleochroism:
Twinning:
Cleavage: Two distinct {100}, less distinct {110}, parallel to length. In cross sections, at right angles
Extinction: Parallel, basal sections remain dark
Alteration: Alters to muscovite
Features: Prismatic crystals, parallel extinction and strong birefringence
Occurrence: Occurs within contact metamorphic limestones with idocrase, diopside and garnet. Also in some gneisses

Physical Properties

Name: Scheelite
Group: Scheelite
Formula: CaWO4
Crystal System: Tetragonal
Color: Colorless, white, grey, brownish, orange-yellow, greenish, purplish
Opacity: Transparent to translucent
Luster: Vitreous to adamantine
Streak: White to yellowish
SGLow: 6.10
SGHigh: 6.10
HardnessLow: 4.5
HardnessHigh: 5
Cleavage: 3
Direction: {101} distinct; {112} interrupted; {001} indistinct
Habit: Crystals octahedral or tabular, often diagonally striated; massive, granular; columnar
Fracture: Subconchoidal to uneven
Other: Occurs mainly within contact metamorphic zones
Comments: Series with powellite; fluoresces bright bluish white to white or yellowish white under SW

Optical Properties

Name: Scheelite
Group: Scheelite
Formula: CaWO4
Crystal System: Tetragonal
Color: Colorless to white. Yellow to orange -yellow when molybdenum bearing
Form: Granular, sometimes tetragonal bipyramids
Relief: High, n>balsam
Birefringence: Moderate, 0.017
2V:
Nalpha or Nord.: 1.920
NBeta or Nextr.: 1.937
NGamma:
Optical Sign: Uniaxial positive
Orientation:
Pleochroism:
Twinning:
Cleavage: Three directions, {101} distinct, {112} poor, {001} indistinct
Extinction: Parallel or symmetrical
Alteration: Isomorphous with powellite
Features: Fluoresces bright bluish white to white or yellowish white under SW
Occurrence: Occurs mainly within contact metamorphic zones

Physical Properties

Name: Schorl (Tourmaline)

Group: Tourmaline

Formula: NaFe3Al6(BO3)3Si6O18(OH)4

Crystal System: Trigonal

Color: Black, brownish black, bluish black

Opacity: Translucent to opaque

Luster: Vitreous to resinous

Streak: Colorless

SGLow: 3.10

SGHigh: 3.25

HardnessLow: 7

HardnessHigh: 7

Cleavage: 2

Direction: {11-20} and {10-11} indistinct

Habit: Crystals short to long prismatic, vertically striated, or acicular to fibrous, thin tablets rare; radiating groups; massive

Fracture: Conchoidal to uneven; brittle

Other: Mainly found within granite pegmatites, granites and high temperature veins

Comments: Series with dravite

Optical Properties

Name: Schorl (Tourmaline)

Group: Tourmaline

Formula: NaFe3Al6(BO3)3Si6O18(OH)4

Crystal System: Trigonal

Color: Grey, slate blue, buff, olive

Form: Prismatic with alongated, hexagonal or triangular outlines in cross section, columnar and fibrous radiating aggregates

Relief: High, n>balsam

Birefringence: Moderate to strong, 0.022-0.040

2V:

Nalpha or Nord.: 1.628-1.658

NBeta or Nextr.: 1.652-1.698

NGamma:

Optical Sign: Uniaxial negative

Orientation: Length fast

Pleochroism: Strong pleochroism

Twinning:

Cleavage: Two {11-20} and {10-11} indistinct

Extinction: Parallel, cross sections remain dark

Alteration:

Features: Elongated and hexagonal outlines, color, pleochroism, parallel extinction and lack of cleavage distinctive.

Occurrence: Mainly found within granite pegmatites, granites and high temperature veins

Physical Properties

Name: Scolecite (Zeolite Group)
Group: Zeolite
Formula: $CaAl_2Si_3O_{10}+3H_2O$
Crystal System: Monoclinic
Color: Colorless, white
Opacity: Transparent to translucent
Luster: Vitreous, silky when fibrous
Streak:
SGLow: 2.27
SGHigh: 2.27
HardnessLow: 5
HardnessHigh: 5
Cleavage: 1
Direction: {110} perfect
Habit: Crystals slender prismatic, vertically striated, divergent groups, commonly twinned; fibrous radiated masses
Fracture: Uneven; brittle
Other: In cavities and veins within lavas, particularly basalts
Comments:

Optical Properties

Name: Scolecite (Zeolite Group)
Group: Zeolite
Formula: $CaAl_2Si_3O_{10}+3H_2O$
Crystal System: Monoclinic
Color: Colorless
Form: Fibrous and clumnar crystal aggregates
Relief: Low, n<balsam
Birefringence: Weak, 0.007
2V: 36"
Nalpha or Nord.: 1.512
NBeta or Nextr.: 1.519
NGamma: 1.519
Optical Sign: Biaxial negative
Orientation: Length fast, r<v strong
Pleochroism:
Twinning: Twinning common
Cleavage: Two directions distinct {110} and at angles 88"
Extinction: Oblique minus 15" - minus 18" in longitudinal sections
Alteration:
Features: Similar to other zeolites, oblique extinction and twinning distinctive
Occurrence: In cavities and veins within lavas, particularly basalts

Optical Properties

Name: Sepiolite

Group: Clay

Formula: 2MgO.3SiO2.nH2O

Crystal System: Monoclinic

Color: Colorless to grey

Form: Fibrous and in aggregates

Relief: Low, n<balsam

Birefringence: Strong, 0.015-0.020

2V: 40"-60"

Nalpha or Nord.: 1.490-1.506

NBeta or Nextr.: 1.505-1.526

NGamma:

Optical Sign: Biaxial negative

Orientation: Length slow

Pleochroism:

Twinning:

Cleavage:

Extinction: Parallel to length

Alteration:

Features: Fibrous aggregates with curved and matted fibre groups.

Occurrence: Contact metamorphic zones as veinlike masses

Physical Properties

Name: Serpentine
Group: Serpentine
Formula: Mg3Si2O5(OH)4
Crystal System: Monoclinic
Color: Yellow, white, grey, green
Opacity: Opaque
Luster: Greasy, waxy, silky
Streak: White
SGLow: 2.55
SGHigh: 2.56
HardnessLow: 2.5
HardnessHigh: 2.5
Cleavage: 1
Direction: {001} perfect
Habit: Fibrous, tabular
Fracture: Splintery, brittle, uneven
Other: A metamorphic mineral principally found in serpentinites
Comments: A group name

Optical Properties

Name: Serpentine
Group: Serpentine
Formula: Mg3Si2O5(OH)4
Crystal System: Monoclinic
Color:
Form:
Relief:
Birefringence:
2V:
Nalpha or Nord.:
NBeta or Nextr.:
NGamma:
Optical Sign:
Orientation:
Pleochroism:
Twinning:
Cleavage:
Extinction:
Alteration:
Features: A group name
Occurrence: A metamorphic mineral principally found in serpentinites

Physical Properties

Name: Siderite

Group: Calcite

Formula: $FeCO_3$

Crystal System: Trigonal

Color: Pale yellowish, pale green, yellowish brown, brown, reddish brown, white

Opacity: Translucent to subtranslucent

Luster: Vitreous, pearly or silky

Streak: White

SGLow: 3.96

SGHigh: 3.96

HardnessLow: 4

HardnessHigh: 4

Cleavage: 1

Direction: {10-11} perfect

Habit: Crystals rhombohedral or tabular, prismatic, scalenohedral; massive, coarse to fine granular; botryoidal, globular

Fracture: Conchoidal to uneven; brittle

Other: Occurs as diagenetic cement in sandstones, in veins and replacement deposits and sedimentary oolitic ironstones

Comments: Series with magnesite and with rhodochrosite

Optical Properties

Name: Siderite

Group: Calcite

Formula: $FeCO_3$

Crystal System: Hexagonal

Color: Colorless to grey, brownish

Form: Anhedral to euhedral fine to coarse aggregates. Oolitic, spherulitic or colloform

Relief: High along long driection, low along short direction

Birefringence: Extreme, 0.234-0.242

2V: 3.96

Nalpha or Nord.: 1.596-1.633

NBeta or Nextr.: 1.830-1.875

NGamma:

Optical Sign: Uniaxial negative

Orientation:

Pleochroism:

Twinning: Polysynthetic parallel to long direction {011_2} occasionally

Cleavage: One perfect {101_1}

Extinction: Symmetrical to cleavage

Alteration:

Features: Similar calcite, dolomite and magnesite. Brown iron staining distinctive. Index of refraction < balsam, others index of refraction > balsam

Occurrence: Occurs as diagenetic cement in sandstones, in veins and replacement deposits and sedimentary oolitic ironstones

Physical Properties

Name: Sillimanite
Group: Sillimanite
Formula: Al2SiO5
Crystal System: Orthorhombic
Color: Colorless, white, grey, yellowish, brownish, greenish, bluish
Opacity: Transparent to translucent
Luster: Vitreous to silky
Streak: Colorless
SGLow: 3.23
SGHigh: 3.27
HardnessLow: 6.5
HardnessHigh: 7.5
Cleavage: 1
Direction: {010} perfect
Habit: Crystals long prismatic, vertically striated, nearly square cross section; massive, fibrous to somewhat columnar
Fracture: Uneven
Other: Widespread within gneisses, schists, slates, hornfelses and other metamorphic rocks
Comments: Trimorphous with andalusite and kyanite

Optical Properties

Name: Sillimanite
Group: Sillimanite
Formula: Al2SiO5
Crystal System: Orthorhombic
Color: Colorless
Form: Small, often minute prismatic crystals and fibrous felted masses. Nearly square in cross section
Relief: High, n>balsam
Birefringence: Moderate, 0.020-0.023
2V: 20"-30"
Nalpha or Nord.: 1.657-1.661
NBeta or Nextr.: 1.658-1.670
NGamma: 1.6777-1.684
Optical Sign: Biaxial positive
Orientation: Length slow, r>v strong
Pleochroism:
Twinning:
Cleavage: One perfect parallel to {010} but not always apparent
Extinction: Parallel in longitudinal sections and symmetrical in cross sections
Alteration:
Features: Similar to andalusite but is length slow, lacks two cleavage at right angles, stronger birefringence.
Occurrence: Widespread within gneisses, schists, slates, hornfelses and other metamorphic rocks

Physical Properties

Name: Smectite
Group: Smectite, Clay
Formula: (0.5Ca,Na)0.7(Al,Mg,Fe)4[Si,Al]8O20](OH)4.nH2O
Crystal System: Monoclinic
Color: White, yellow-green
Opacity: Opaque
Luster: Non metaliic
Streak:
SGLow: 2
SGHigh: 3
HardnessLow: 1
HardnessHigh: 2
Cleavage: 1
Direction: {001} perfect
Habit: Fine grained aggregates, vermiform, lamellar or spherulitic
Fracture: Plates
Other: Usually found as a diagenetic clay mineral in sedimentary beds. Alteration product oof volcanic ash and tuff
Comments: Major group of clay minerals. Synonomous with montmorillonite

Optical Properties

Name: Smectite
Group: Smectite, Clay
Formula: (0.5Ca,Na)0.7(Al,Mg,Fe)4[Si,Al]8O20](OH)4.nH2O
Crystal System: Monoclinic
Color:
Form: Fine grained aggregates, vermiform, lamellar or spherulitic, plates
Relief:
Birefringence:
2V:
Nalpha or Nord.:
NBeta or Nextr.:
NGamma:
Optical Sign:
Orientation:
Pleochroism:
Twinning:
Cleavage: One {001} perfect
Extinction:
Alteration:
Features: Major group of clay minerals. Synonomous with montmorillonite
Occurrence: Usually found as a diagenetic clay mineral in sedimentary beds. Alteration product of volcanic ash and tuff

Physical Properties

Name: Sodalite
Group: Feldspathoid
Formula: Na4Al3(SiO4)3Cl
Crystal System: Isometric, cubic
Color: Grey, green, colorless, white or blue
Opacity: Transparent to translucent
Luster: Vitreous, greasy
Streak: White
SGLow: 2.2
SGHigh: 2.3
HardnessLow: 5.5
HardnessHigh: 6.0
Cleavage: Poor, six directions
Direction:
Habit: Crystals rare as dodecahedrons. Usually compact, nodular or disseminated grains
Fracture: Uneven to conchoidal
Other: Occurs in alkali igneous rocks, such as nepheline syenites.
Comments:

Optical Properties

Name: Sodalite
Group: Feldspathoid
Formula: 3NaAlSiO4.NaCl
Crystal System: Isometric
Color: Colorless to grey, often with dark borders
Form: Six sided cros sections of dodecahedrons and anhedral crystals
Relief: Fair, n<balsam
Birefringence: None
2V:
Nalpha or Nord.: 1.483-1.487
NBeta or Nextr.:
NGamma:
Optical Sign:
Orientation:
Pleochroism:
Twinning:
Cleavage: One imperfect {110}
Extinction: Isotropic
Alteration: Easily altered to zeolites
Features: Staining may be required
Occurrence: Characteristic of alkali igneous rocks, such as syenites and trachytes.

Physical Properties

Name: Spessartite (Spessartine)

Group: Garnet

Formula: Mn3Al2(SiO4)3

Crystal System: Cubic

Color: Red, reddish orange, yellowish brown, reddish brown to brown

Opacity: Transparent to translucent

Luster: Vitreous

Streak: White

SGLow: 3.8

SGHigh: 4.25

HardnessLow: 7

HardnessHigh: 7.5

Cleavage: 1

Direction: {110} parting may be distinct

Habit: Crystals dodecahedrons or trapezohedrons, faces often striated; embedded grains; massive, compact

Fracture:

Other: Occurs in pegmatites and metamorphic schists and quartzites

Comments: Series with almandine

Optical Properties

Name: Spessartite (Spessartine)

Group: Garnet

Formula: Mn3Al2(SiO4)3

Crystal System: Isometric, cubic

Color: Colorless to pale red, pale to dark brown, greenish

Form: Euhedral dodecahedrons in six sided trapezohedrons in eigth sided cross sections. Plygonal grains and aggregates.

Relief: V.high, n>balsam

Birefringence: Isotropic but may show weak birefringence

2V:

Nalpha or Nord.: 1.792-1.820

NBeta or Nextr.:

NGamma:

Optical Sign:

Orientation:

Pleochroism:

Twinning:

Cleavage: Parting parallel to {110}, irregular fractures

Extinction:

Alteration:

Features: Similar to spinel which is octohedral. Determination of RI will differentiate garnets.

Occurrence: Occurs in pegmatites and metamorphic schists and quartzites

Physical Properties

Name: Sphene (Titanite)
Group: Garnet
Formula: CaTiSiO5
Crystal System: Monoclinic
Color: Colorless, yellow, green, grey, brown, rose-red, black, pinkish red
Opacity: Transparent to nearly opaque
Luster: Adamantine to resinous
Streak: White
SGLow: 3.45
SGHigh: 3.55
HardnessLow: 5
HardnessHigh: 5.5
Cleavage: 2
Direction: {110} distinct, {221} parting
Habit: Crystals wedge-shaped or prismatic, twinned; massive, compact or lamellar
Fracture:
Other: Widespread accessory mineral in igneous and metamorphic rocks
Comments:

Optical Properties

Name: Sphene
Group: Garnet
Formula: CaTiSO5
Crystal System: Monoclinic
Color: Colorless to neutral
Form: Usually euhedral with acute rhombic cross section. Also as irregular grains
Relief: V.high, n>balsam
Birefringence: Extreme, 0.092-0.141
2V: 23"-50"
Nalpha or Nord.: 1.887-1.913
NBeta or Nextr.: 1.894-1.921
NGamma: 1.973-2.054
Optical Sign: Biaxial positive
Orientation: r>v strong
Pleochroism: Thick sections may be pleochroic from nearly colorless-pale yellow or greenish-yellow to red brown
Twinning: Simple twinning along {100} and polysynthetic twinning along {221) infrequent
Cleavage: Two {110} distinct, {221} parting. Often with parting parallel to {221}. Shows pseudo rhombic parting
Extinction: Rhombic sections have symmetrical extinction. Frequently with incomplete extinction because of strong dispersion
Alteration:
Features: Acute rhombic cross sections and color distinctive. Similar to monazite but has lower birefringence.
Occurrence: Widely distributed accessory mineral igneous and metamorphic rocks.

Physical Properties

Name: Spinel
Group: Spinel
Formula: MgAl2O4
Crystal System: Cubic
Color: Red, blue, green, brown, black
Opacity: Transparent to opaque
Luster: Vitreous to nearly dull
Streak: White
SGLow: 3.58
SGHigh: 3.58
HardnessLow: 7.5
HardnessHigh: 8
Cleavage: 1
Direction: {111} parting indistinct
Habit: Crystals octahedral, may be modified, or cubic or dodecahedral; massive, compact to coarse granular; rounded grains
Fracture: Conchoidal or uneven; brittle
Other: Accessory mineral within igneous and metamorphic rocks.
Comments: Series with magnesiochromite with gahnite and with hercynite

Optical Properties

Name: Spinel
Group: Spinel
Formula: MgAl2O4
Crystal System: Isometric, cubic
Color: Colorless to red, green (pleonaste), olive green or brown (picotite)
Form: Euhadral or subhedral octohedra with rhombic sections. Also equant grains
Relief: High, n>balsam
Birefringence: Isotropic
2V:
Nalpha or Nord.: 1.72-1.78
NBeta or Nextr.:
NGamma:
Optical Sign:
Orientation:
Pleochroism:
Twinning: According to spinel law {111}
Cleavage: one 111} parting indistinct. Imperfect octohedral
Extinction:
Alteration:
Features: Octohedral form, color, birefringence.
Occurrence: Accessory mineral within igneous and metamorphic rocks.

Physical Properties

Name: Spodumene
Group: Pyroxene
Formula: LiAlSi2O6
Crystal System: Monoclinic
Color: Colorless, white, grey, yellowish, greenish, emerald-green, pink, violet
Opacity: Transparent to translucent
Luster: Vitreous to dull
Streak: White
SGLow: 3.0
SGHigh: 3.2
HardnessLow: 6.5
HardnessHigh: 7.5
Cleavage: 3
Direction: {110} perfect, {010} and {100} parting
Habit: Crystals prismatic, may be flattened, striated vertically, large; massive, cleavable
Fracture:
Other: A lithium mineral found mainly in granite pegmatites with other lithium minerals
Comments:

Optical Properties

Name: Spodumene
Group: Pyroxene
Formula: LiAlSi2O6
Crystal System: Monoclinic
Color: Colorless. Kunzite (amethystine) and hiddenite (greenish).
Form: Euhedral tabular crystals, elongated along {001}
Relief: High, n>balsam
Birefringence: Moderate, 0.013-0.027
2V: 54"-69"
Nalpha or Nord.: 1.651-1.668
NBeta or Nextr.: 1.665-1.675
NGamma: 1.677-1.681
Optical Sign: Biaxial positive
Orientation: Length slow, r<v
Pleochroism: Thick colored sections are pleochroic
Twinning: Simple twins along {100} common
Cleavage: Three perfect parallel to {110} and (11_0) at 93". arting parallel to {100}, often more distinct than cleavage.
Extinction: Oblique 23"-27" in longitudinal sections, parallel or symmetrical in cross sections
Alteration: Alters to a mixture of albite and muscovite known as cymatolite
Features: Similar to diopside, but with smaller extinction angle. Occurrence distinctive.
Occurrence: A lithium mineral found mainly in granite pegmatites with other lithium minerals

Physical Properties

Name: Staurolite
Group: Garnet
Formula: (Fe,Mg,Zn)2Al9(Si,Al)4O22(OH)2
Crystal System: Monoclinic ps. orth.
Color: Dark brown, reddish brown, yellowish brown, brownish black
Opacity: Translucent to nearly opaque
Luster: Vitreous to resinous
Streak: Colorless, greyish
SGLow: 3.65
SGHigh: 3.83
HardnessLow: 7
HardnessHigh: 7.5
Cleavage: 1
Direction: {010} distinct
Habit: Crystals short prismatic, rough surfaces, cruciform twins
Fracture: Subconchoidal to uneven; brittle
Other: Occurs in metamorphic schists and gneisses
Comments:

Optical Properties

Name: Staurolite
Group: Garnet
Formula: (Fe,Mg,Zn)2Al9(Si,Al)4O22(OH)2
Crystal System: Monoclinic ps. orth.
Color: Pale yellow
Form: Prismatic euhedral, rectangular and six sided cross sections. Large crystals 1cm long common
Relief: High, n>balsam
Birefringence: Weak, 0.010-0.015
2V: 80"-88"
Nalpha or Nord.: 1.736-1.747
NBeta or Nextr.: 1.741-1.754
NGamma: 1.746-1.762
Optical Sign: Biaxial positive
Orientation: Length slow, r>v weak
Pleochroism: Strong pleochroism from nearly colorless-yellow brown
Twinning: Penetration twins with {023} and {232} frequent
Cleavage: One non distinct, parallel to {010}
Extinction: Parallel in longitudinal sections, symmetrical in cross sections
Alteration:
Features: Color and pleochroism are distinctive
Occurrence: Occurs in metamorphic schists and gneisses

Physical Properties

Name: Stilbite (Zeolite)
Group: Zeolite
Formula: NaCa2Al5Si13O36+14H2O
Crystal System: Monoclinic
Color: White, grey, yellowish, pink, reddish, orange, light to dark brown
Opacity: Transparent to translucent
Luster: Vitreous, pearly on cleavage
Streak: Colorless
SGLow: 2.09
SGHigh: 2.20
HardnessLow: 3.5
HardnessHigh: 4
Cleavage: 2
Direction: {010} perfect, {100} distinct
Habit: Single crystals rare; bladed aggregates or twinned crystals widening at terminations forming hourglass or sheaf-like forms
Fracture: Uneven; brittle
Other: In cavities and veins within lavas, particularly basalts
Comments: Also orth.

Optical Properties

Name: Stilbite (Zeolite)
Group: Zeolite
Formula: NaCa2Al5Si13O36+14H2O
Crystal System: Monoclinic
Color: Colorless
Form: Usually in sheaflike aggregates
Relief: Low, n<balsam
Birefringence: Weak 0.006-0.008
2V: 33"
Nalpha or Nord.: 1.494-1.500
NBeta or Nextr.: 1.498-1.504
NGamma: 1.500-1.508
Optical Sign: Biaxial negative
Orientation: r<v
Pleochroism:
Twinning: Simple twins along {001} common
Cleavage: One good, {010}
Extinction: Parallel to cleavage, usually wavy
Alteration:
Features: Cleavage, biaxial positive, twinning.
Occurrence: In cavities and veins within lavas, particularly basalts

Physical Properties

Name: Stilpnomelane

Group: Stilpnomelane

Formula: K(Fe,Al)10Si12O30(OH)12

Crystal System: Monoclinic and triclinic

Color: Black, deep reddish brown, golden brown, dark green

Opacity: Translucent to nearly opaque

Luster: Pearly to subvitreous

Streak:

SGLow: 2.59

SGHigh: 2.96

HardnessLow: 3

HardnessHigh: 3

Cleavage: 2

Direction: {001} perfect, {010} imperfect

Habit: Foliated plates; coatings, fibrous or velvety

Fracture: Brittle

Other: Occurs within hydrothermal veins and nepheline-syenite pegmaities

Comments:

Optical Properties

Name: Stilpnomelane

Group: Brittle Mica

Formula: K(Fe,Al)10Si12O30(OH)12

Crystal System: Monoclinic and triclinic

Color: Brown and yellow to green

Form: Usually micaceous masses

Relief: Moderate, n>balsam

Birefringence: Moderate to strong, 0.030-0.119

2V: 0"

Nalpha or Nord.: 1.612-1.634

NBeta or Nextr.: 1.700-1.745

NGamma: 1.700-1.745

Optical Sign: Pseudo Uniaxial negative

Orientation:

Pleochroism: Pleochroic

Twinning: Plysynthetic according to mica law

Cleavage: One perfect {001}

Extinction: Almost parallel

Alteration:

Features: Similar to biotite but pseudo uniaxial figure and higher RI are distinctive.

Occurrence: Occurs within quartz-adularia hydrothermal veins and nepheline-syenite pegmaities. Probably a hydrothermal alteration product.

Physical Properties

Name: Talc
Group: Talc
Formula: Mg3Si4O10(OH)2
Crystal System: Monoclinic and triclinic
Color: Pale green to dark green, greenish grey, white, silvery white, grey, brownish
Opacity: Translucent
Luster: Pearly or dull
Streak: White
SGLow: 2.58
SGHigh: 2.83
HardnessLow: 1
HardnessHigh: 1
Cleavage: 1
Direction: {001} perfect
Habit: Crystals thin tabular, to 1 cm; massive, fine-grained compact, foliated or fibrous; globular stellate groups
Fracture: Laminae flexible, inelastic
Other: Found within metamorphic rocks, especially schists. Alteration product of actinolite and tremolite
Comments:

Optical Properties

Name: Talc
Group: Brittle Mica
Formula: Mg3Si4O10(OH)2
Crystal System: Monoclinic and triclinic
Color: Colorless
Form: Cleavage masses in coarse to fine platy or fibrous aggregates. Euhedral crystals unknown
Relief: Fair, n>balsam
Birefringence: Very strong, 0.030-0.050. Low, first order grey parallel to cleavage
2V: 6"=30"
Nalpha or Nord.: 1.538-1.545
NBeta or Nextr.: 1.575-1.590
NGamma: 1.575-1.590
Optical Sign: Biaxial negative
Orientation: Length slow, r>v distinct
Pleochroism:
Twinning:
Cleavage: One perfect {001}
Extinction: Parallel upto 3"
Alteration:
Features: Similar to muscovite and pyrophyllite, but physical examination of sample will usually distinguish.
Occurrence: Found within metamorphic rocks, especially schists. Alteration product of actinolite and tremolite

Physical Properties

Name: Thomsonite (Zeolite Group)

Group: Zeolite

Formula: NaCa2Al5Si5O20+6H2O

Crystal System: Orthorhombic

Color: Colorless, white,yellowish, pink, greenish

Opacity: Transparent to translucent

Luster: Vitreous to somewhat pearly

Streak: Colorless

SGLow: 2.25

SGHigh: 2.40

HardnessLow: 5

HardnessHigh: 5.5

Cleavage: 2

Direction: {010} perfect; {100} distinct

Habit: Crystals prismatic, acicular or blade-like, striated vertically, rare; lamellar or radiating aggregates; compact

Fracture: Subconchoidal to uneven; brittle

Other: In cavities and veins within lavas, particularly basalts

Comments:

Optical Properties

Name: Thomsonite (Zeolite Group)

Group: Zeolite

Formula: NaCa2Al5Si5O20+6H2O

Crystal System: Orthorhombic

Color: Colorless

Form: Fibrous or columnar aggregates

Relief: Low, n<balsam

Birefringence: Weak to moderate, 0.006-0.012

2V: 44"-55"

Nalpha or Nord.: 1.512-1.530

NBeta or Nextr.: 1.513-1.532

NGamma: 1.518-1.542

Optical Sign: Biaxial positive

Orientation: Some length slow, some length fast, r>v strong

Pleochroism:

Twinning:

Cleavage: One direction {010}

Extinction: Parallel

Alteration:

Features: Similar to other zeolites

Occurrence: In cavities and veins within lavas, particularly basalts

Physical Properties

Name: Titanite (Sphene)
Group: Garnet
Formula: CaTiSiO5
Crystal System: Monoclinic
Color: Colorless, yellow, green, grey, brown, rose-red, black, pinkish red
Opacity: Transparent to nearly opaque
Luster: Adamantine to resinous
Streak: White
SGLow: 3.45
SGHigh: 3.55
HardnessLow: 5
HardnessHigh: 5.5
Cleavage: 2
Direction: {110} distinct, {221} parting
Habit: Crystals wedge-shaped or prismatic, twinned; massive, compact or lamellar
Fracture:
Other: Widespread accessory mineral in igneous and metamorphic rocks
Comments:

Optical Properties

Name: Titanite (Sphene)
Group: Garnet
Formula: CaTiSO5
Crystal System: Monoclinic
Color: Colorless to neutral
Form: Usually euhedral with acute rhombic cross section. Also as irregular grains
Relief: V.high, n>balsam
Birefringence: Extreme, 0.092-0.141
2V: 23"-50"
Nalpha or Nord.: 1.887-1.913
NBeta or Nextr.: 1.894-1.921
NGamma: 1.973-2.054
Optical Sign: Biaxial positive
Orientation: r>v strong
Pleochroism: Thick sections may be pleochroic from nearly colorless-pale yellow or greenish-yellow to red brown
Twinning: Simple twinning along {100} and polysynthetic twinning along {221) infrequent
Cleavage: Two {110} distinct, {221} parting. Often with parting parallel to {221}. Shows pseudo rhombic parting
Extinction: Rhombic sections have symmetrical extinction. Frequently with incomplete extinction because of strong dispersion
Alteration:
Features: Acute rhombic cross sections and color distinctive. Similar to monazite but has lower birefringence.
Occurrence: Widely distributed accessory mineral igneous and metamorphic rocks.

Physical Properties

Name: Topaz
Group: Topaz
Formula: Al2SiO4(F,OH)2
Crystal System: Orthorhombic
Color: Colorless, white, grey, bluish, greenish, pink, yellow-brown to orange, reddish
Opacity: Transparent to translucent
Luster: Vitreous
Streak: Colorless
SGLow: 3.49
SGHigh: 3.57
HardnessLow: 8
HardnessHigh: 8
Cleavage: 1
Direction: {001} perfect
Habit: Crystals short to long prismatic, may be highly modified, to large size; massive, coarse to fine granular; columnar
Fracture: Subconchoidal to uneven; brittle
Other: Occurs within granite pegmatites and high temperature veins
Comments:

Optical Properties

Name: Topaz
Group: Sillimanite
Formula: Al2SiO4(F,OH)2
Crystal System: Orthorhombic
Color: Colorless
Form: Short, prismatic euhedral crystals, anhedral grains and columnar aggregates
Relief: High, n>balsam
Birefringence: Weak, 0.009-0.010
2V: 48"-65"
Nalpha or Nord.: 1.607-1.629
NBeta or Nextr.: 1.610-1.631
NGamma: 1.617-1.638
Optical Sign: Biaxial positive
Orientation: Length fast, r>v distinct
Pleochroism:
Twinning:
Cleavage: One perfect parllel to {001}
Extinction: Parallel in longitudinal sections and symmetrical in cross sections
Alteration: Alters to muscovite and sericite
Features: Similar to quartz but has higher relief, perfect cleavage and is biaxial.
Occurrence: Occurs within granite pegmatites and high temperature veins

Descriptive Handbook of Rock Forming Minerals

Physical Properties

Name: Tremolite

Group: Amphibole

Formula: Ca2(Mg,Fe)5Si8O22(OH)2

Crystal System: Monoclinic

Color: Colorless, white, grey, pale greenish, pink, brown

Opacity: Transparent to translucent

Luster: Vitreous

Streak: Colorless

SGLow: 2.9

SGHigh: 3.2

HardnessLow: 5

HardnessHigh: 6

Cleavage: 2

Direction: {110} good, {100} parting

Habit: Crystals long-bladed or short and stout; fibrous or thin-columnar aggregates; massive, fibrous or granular

Fracture: Subconchoidal to uneven; brittle, some compact varieties tough

Other: Widespread within metamorphic rocks, especially schists and meta-igneous rocks. Alteration product of pyroxenes and olivine. Also in metamorphosed limestones and skarns

Comments: Series with actinolite and ferro-actinolite

Optical Properties

Name: Tremolite

Group: Amphibole

Formula: Ca2(Mg,Fe)5Si8O22(OH)2

Crystal System: Monoclinic

Color: Colorless to pale green

Form: Long prismatic and columnar to fibrous aggregates

Relief: High, n>balsam

Birefringence: Moderate to strong, 0.022-0.027

2V: 79"-85"

Nalpha or Nord.: 1.600-1.628

NBeta or Nextr.: 1.613-1.644

NGamma: 1.625-1.655

Optical Sign: Biaxial negative

Orientation: Length slow, r<v

Pleochroism: Green varieties show weak pleochroism

Twinning: Polysynthetic twinning along {001} infrequent. Twinning along {100} frequent

Cleavage: In two directions {110} at 56" and 124", parallel in longitudinal sections

Extinction: Oblique 10"-20" in longitudinal sections, symmetrical in cross sections

Alteration: Alters to talc

Features: Extinction angle amphibole cross section characteristic

Occurrence: Widespread within metamorphic rocks, especially schists and meta-igneous rocks. Alteration product of pyroxenes and olivine. Also in metamorphosed limestones and skarns

Physical Properties

Name: Tridymite
Group: Quartz
Formula: SiO2
Crystal System: Monoclinic ps. hex.
Color: Colorless to white
Opacity: Transparent
Luster: Vitreous, sometimes pearly on {001}
Streak: Colorless
SGLow: 2.26
SGHigh: 2.27
HardnessLow: 7
HardnessHigh: 7
Cleavage: 0
Direction:
Habit: Crystals tabular, pseudohexagonal
Fracture: Conchoidal; brittle
Other: Usually occurs within cavities of volcanic igneous rocks
Comments: Polymorphous with coesite, cristobalite, quartz and stishovite

Optical Properties

Name: Tridymite
Group: Quartz
Formula: SiO2
Crystal System: Orthorhombic
Color: Colorless
Form: Minute, euhedral hexaginal and tabular crystals, often twinned. Crystalline aggregates
Relief: Moderate, n<balsam
Birefringence: Weak, 0.004
2V: 35"
Nalpha or Nord.: 1.469
NBeta or Nextr.: 1.469
NGamma: 1.473
Optical Sign: Biaxial positive
Orientation:
Pleochroism:
Twinning: Wege shaped twins
Cleavage: None
Extinction: Wavy to even
Alteration: Polymorphous with quartz and crystabolite
Features: Similar to crystabollite, wedge shaped twinning distinctive, may need to determine RI to differentiate
Occurrence: Mostly in cavities of volcanic rocks

Physical Properties

Name: Uraninite
Group: Uraninite
Formula: UO2
Crystal System: Cubic
Color: Black to brownish or greyish black
Opacity: Opaque
Luster: Submetallic, pitchy, greasy
Streak: Black to brownish black, greyish
SGLow: 7.5
SGHigh: 10.0
HardnessLow: 5
HardnessHigh: 6
Cleavage: 1
Direction: Octahedral
Habit: Crystals cubic, cubo-octahedral or modified octahedrons or dodecahedrons; massive, dense to granular; botryoidal
Fracture: Conchoidal to uneven; brittle
Other: Occurs in pegmatites, veins and hydrothermal replacement deposits
Comments: Usually partially oxidized; series with thorianite

Optical Properties

Name: Uraninite
Group: Uraninite
Formula: UO2
Crystal System: Isometric, cubic
Color: Black to brownish black, greyish
Form: Usually anhedral, botryoidal, colloform
Relief: V. high
Birefringence: Opaque
2V:
Nalpha or Nord.:
NBeta or Nextr.:
NGamma:
Optical Sign:
Orientation:
Pleochroism:
Twinning:
Cleavage: One octahedral
Extinction:
Alteration:
Features: Strongly radioactive. Usually partially oxidized; series with thorianite. View in reflected light on polished section.
Occurrence: Occurs in pegmatites, veins and hydrothermal replacement deposits

Physical Properties

Name: Uvarovite
Group: Garnet
Formula: $Ca_3Cr_2(SiO_4)_3$
Crystal System: Cubic
Color: Emerald-green
Opacity: Transparent to translucent
Luster: Vitreous
Streak: White
SGLow: 3.4
SGHigh: 3.8
HardnessLow: 6.5
HardnessHigh: 7
Cleavage: 0
Direction:
Habit: Crystals dodecahedrons or trapezohedrons; massive, coarse granular; embedded grains
Fracture:
Other: Rare within contact metamorphic zones
Comments: Series with grossular

Optical Properties

Name: Uvarovite (Garnet)
Group: Garnet
Formula: $Ca_2Cr_2(SiO_4)_3$
Crystal System: Isometric, cubic
Color: Colorless to pale red, pale to dark brown, often zoned, greenish
Form: Euhedral dodecahedrons in six sided trapezohedrons in eigth sided cross sections. Plygonal grains and aggregates.
Relief: V.high, n>balsam
Birefringence: Isotropic, but may show weak birefringence
2V:
Nalpha or Nord.: 1.838-1.870
NBeta or Nextr.:
NGamma:
Optical Sign:
Orientation:
Pleochroism:
Twinning:
Cleavage: Parting parallel to {110}, irregular fractures
Extinction:
Alteration:
Features: Similar to spinel, which is octohedral. Determination of RI will differentiate garnets.
Occurrence: Rare, secondary in chromite and in some contact metamorphic zones.

Physical Properties

Name: Vermiculite
Group: Vermiculite, Clay
Formula: $(Mg,Fe,Al)_3(Al,Si)_4O_{10}(OH)_2+4H_2O$
Crystal System: Monoclinic
Color: Greyish-brownish
Opacity: Opaque
Luster: Non metallic
Streak:
SGLow: 2.756
SGHigh: 2.756
HardnessLow: 1
HardnessHigh: 2
Cleavage:
Direction:
Habit:
Fracture:
Other: A weathering product of igneous and metamorphic rocks and in clay beds. Also common as a diagenetic clay mineral in sedimentary beds
Comments: A group of silicates with the general formula above. A group of clay minerals derived from alteration of mica

Optical Properties

Name: Vermiculite
Group: Vermiculite, Clay
Formula: $(Mg,Fe,Al)_3(Al,Si)_4O_{10}(OH)_2+4H_2O$
Crystal System: Monoclinic
Color:
Form:
Relief:
Birefringence:
2V:
Nalpha or Nord.:
NBeta or Nextr.:
NGamma:
Optical Sign:
Orientation:
Pleochroism:
Twinning:
Cleavage:
Extinction:
Alteration:
Features: A group of silicates with the general formula above. A group of clay minerals derived from alteration of mica
Occurrence: A weathering product of igneous and metamorphic rocks and in clay beds. Also common as a diagenetic clay mineral in sedimentary beds

Physical Properties

Name: Vesuvianite (Idocrase)
Group: Garnet
Formula: Ca10Mg2Al4(SiO4)5(Si2O7)2(OH)4
Crystal System: Tetragonal
Color: Green, brown, yellow, white, red, blue, purple
Opacity: Transparent to translucent
Luster: Vitreous, resinous
Streak: White
SGLow: 3.33
SGHigh: 3.45
HardnessLow: 6
HardnessHigh: 7
Cleavage: 3
Direction: {001}, {100} and {110} indistinct
Habit: Crystals short prismatic, sometimes pyramidal; massive, granular, columnar or cryptocrystalline
Fracture: Conchoidal; brittle
Other: Occurs predominantly in contact metamorphic zones
Comments:

Optical Properties

Name: Vesuvianite (Idocrase)
Group: Garnet
Formula: Ca10Mg2Al4(SiO4)5(Si2O7)2(OH)4
Crystal System: Tetragonal
Color: Colorless to neutral
Form: Euhedral prismatic crystals, anhedral with polygonal outlines, columnar aggregates and fine grained aggregates
Relief: High, n>balsam
Birefringence: Weak, 0.004-0.006
2V:
Nalpha or Nord.: 1.701-1.726
NBeta or Nextr.: 1.705-1.732
NGamma:
Optical Sign: Uniaxial negative
Orientation: Length fast,
Pleochroism: Thick sections may be pleochroic
Twinning:
Cleavage: Three {001}, {100} and {110} indistinct
Extinction: Parallel
Alteration:
Features: Crystal form, low birefringence, uniaxial positive
Occurrence: Occurs predominantly in contact metamorphic zones

Physical Properties

Name: Volcanic Glass (Obsidian)
Group: Glass
Formula:
Crystal System:
Color:
Opacity:
Luster:
Streak: Colorless
SGLow:
SGHigh:
HardnessLow:
HardnessHigh:
Cleavage:
Direction:
Habit:
Fracture:
Other: Widespread in volcanic lavas
Comments:

Optical Properties

Name: Volcanic Glass
Group: Mineraloid
Formula: SiO_2
Crystal System: Amorphous
Color: Colorless
Form: Amorphous silica glass, may be banded or show flow structure, vesicular, perlitic, often contains spherulites, microlites, crystallites, microphenocrysts and phenocrysts
Relief: Low to moderate, n<balsam
Birefringence: None, may show weak birefringence
2V:
Nalpha or Nord.: 1.458-1.462
NBeta or Nextr.:
NGamma:
Optical Sign:
Orientation:
Pleochroism:
Twinning:
Cleavage: None
Extinction:
Alteration: Devitified, feldspars, trydimite, cristabolite or monmorilonite from devitrification
Features: Isotropism, low relief and birefringence. Amorphous silica glass, may be banded or show flow structure, vesicular, perlitic, often contains spherulites, microlites, crystallites, microphenocrysts and phenocrysts.
Occurrence: Occurs frequently in the groundmass of volcanic igneous rocks. Often occurs as a rock type, such as obsidian, pumice, perlite or pitchstone. Usually rhyolitic in composition.

Physical Properties

Name: Wollastonite

Group: Pyroxene

Formula: CaSiO3

Crystal System: Triclinic

Color: White to greyish, colorless or very pale green

Opacity: Transparent to translucent

Luster: Vitreous to pearly

Streak: White

SGLow: 2.87

SGHigh: 3.09

HardnessLow: 4.5

HardnessHigh: 5

Cleavage: 3

Direction: {100} perfect, {001} and {-102} good

Habit: Crystals tabular; massive, cleavable to fibrous or granular and compact

Fracture: Splintery

Other: Charcateristic of contact metamorphic zones and in some schists and gneisses

Comments: Trimorphous with wollastonite-2M and wollastonite-7T; sometimes fluoresces, luster somewhat silky if fibrous

Optical Properties

Name: Wollastonite

Group: Pyroxene

Formula: CaSiO3

Crystal System: Triclinic

Color: Colorless

Form: Columnar or fibrous aggregates, almost rectangular cross sections

Relief: High, n>balsam

Birefringence: Weak, 0.014

2V: 39"

Nalpha or Nord.: 1.620

NBeta or Nextr.: 1.632

NGamma: 1.634

Optical Sign: Biaxial negative

Orientation: Length slow or length fast, r>v weak

Pleochroism:

Twinning: Simple twins along {100} rare

Cleavage: Three perfect parallel to {100}, distinct {001} and {1_02}, imperfect {101} and 1_01}

Extinction: Parallel or almost parallel in longitudinal sections, oblique in cross sections

Alteration:

Features: Similar to tremolite, which has amphibole cross section. Cleavage and cross section distinctive

Occurrence: Charcateristic of contact metamorphic zones and in some schists and gneisses

Descriptive Handbook of Rock Forming Minerals

Physical Properties

Name: Zircon
Group: Garnet
Formula: ZrSiO4
Crystal System: Tetragonal
Color: Colorless, brown, green, grey, yellow, red
Opacity: Transparent
Luster: Vitreous to adamantine
Streak: Colorless
SGLow: 4.6
SGHigh: 4.7
HardnessLow: 7.5
HardnessHigh: 7.5
Cleavage: 2
Direction: {110} imperfect; {111} poor
Habit: Crystals short to long prismatic, or dipyramidal; sheaf-like or radial-fibrous aggregates; irregular grains
Fracture: Uneven; conchoidal when metamict; brittle
Other: Widespread mineral in igneous rocks, especially granites where it is mostly an accessory mineral. Widespread detrital mineral
Comments: Transparent to opaque if metamict

Optical Properties

Name: Zircon
Group: Garnet
Formula: ZrSiO4
Crystal System: Tetragonal
Color: Colorless to pale colors
Form: Minute, short prismatic crystals, surrounded by pleochroic haloes
Relief: V.high, n>balsam
Birefringence: Very strong, 0.060-0.062
2V:
Nalpha or Nord.: 1.925-1.931
NBeta or Nextr.: 1.985-1.993
NGamma:
Optical Sign: Uniaxial positive
Orientation: Length slow
Pleochroism:
Twinning:
Cleavage: Two {110} imperfect; {111} poor, not usually visible
Extinction: Parallel
Alteration: Metamict alteration to crytolite, an amorphous mineraloid
Features: High relief and cross section distinctive
Occurrence: Widespread mineral in igneous rocks, especially granites where it is mostly an accessory mineral. Widespread detrital mineral

Physical Properties

Name: Zoisite

Group: Epidote

Formula: Ca2Al3(SiO4)3(OH)

Crystal System: Orthorhombic

Color: Grey, white, greenish grey, greenish brown, green, pink, colorless to blue to purple

Opacity: Transparent to translucent

Luster: Vitreous, may be pearly on cleavage

Streak: Colorless

SGLow: 3.55

SGHigh: 3.55

HardnessLow: 6.5

HardnessHigh: 7

Cleavage: 2

Direction: {100} perfect, {010} imperfect

Habit: Crystals prismatic, deeply striated vertically; massive, compact or columnar

Fracture: Conchoidal to uneven; brittle

Other: Rarely occurs in metamorphic rocks

Comments: Dimorphous with clinozoisite

Optical Properties

Name: Zoisite

Group: Epidote

Formula: Ca2Al3(SiO4)3(OH)

Crystal System: Orthorhombic

Color: Colorless, maganian variety is pink

Form: Clumnar aggregates, euhedral prismatic with rectangular outlines infrequent

Relief: High, n>balsam

Birefringence: Weak to moderate, 0.006-0.018. A second variety shows anomolous deep blue

2V: 30"-60"

Nalpha or Nord.: 1.696-1.700

NBeta or Nextr.: 1.696-1.703

NGamma: 1.702-1.718

Optical Sign: Biaxial positive

Orientation: 1. Length fast or 2. length slow 1. r<v distinct 2. r>v distinct

Pleochroism: Manganian zoisite is pink and pleochroic

Twinning: Polysynthetic infrequent

Cleavage: One perfect along {010}

Extinction: Parallel, remains dark in cross sections

Alteration:

Features: Similar to clinozoisite but has 1. normal birefringence colors, 2. anomolous birefringence

Occurrence: Rarely occurs in metamorphic rocks

Alphabetical Index of Rock Forming Minerals

Acmite (Aegerine)	Chlorite
Actinolite	Chloritoid
Aegerine Augite	Chondrodite
Albite	Chromite
Allanite-[Ce]	Chrysotile
Allanite-[Y]	Clinochlore
Almandine	Clinozoisite
Alunite	Colemantite
Amphibole Group	Cordierite
Analcite (Analcime)	Corundum
Andalusite (Chiastolite)	Cristabolite
Andesine	Crossite
Andradite	Cummingtonite
Anhydrite	Diamond
Anorthite	Diaspore
Anthophyllite	Dickite
Antigorite	Diopside
Apatite	Dolomite
Aragonite	Dravite (Tourmailine Group)
Arfvedsonite	Dumortierite
Augite	Elbaite (Tourmaline Group)
Axinite	Enstatite
Baryte (Barite)	Epidote
Beryl	Erionite (Zeolite Group)
Biotite	Fayalite (Olivine)
Boehmite (Bauxite Group)	Ferrohornblende (Barkevikite)
Boracite	Fluorapatite (Apatite)
Borax	Fluorite
Bronzite	Forsterite (Olivine)
Brucite	Fuschite
Bytownite	Gibbsite (Bauxite)
Calcite	Glaucophane
Cancrinite	Goethite
Carnallite	Graphite
Carnotite	Grossular (Grossularite)
Cassiterite	Grunerite
Celestite	Gypsum
Chabazite	Halite
Chalcedony	Halloysite
Chamosite	Hauyne

Alphabetical Index of Rock Forming Minerals cont...

Hedenbergite	Perovskite
Hematite	Phlogopite
Heulandite	Piedmontite
Hornblende	Plagioclase Group
Humite	Prehnite
Hypersthene	Pumpellyite-[Fe]
Illite	Pumpellyite-[Mg]
Jadeite	Pyromorphite
Jarosite	Pyrope
Kaolinite	Pyrophyllite
Kernite	Pyroxene Group
Kyanite	Quartz
Labradorite	Rhodonite
Laumontite (Zeolite)	Riebeckite (Crocidolite)
Lawsonite	Rubellite (Tourmaline)
Lazulite	Rutile
Lepidolite	Sanidine
Leucite	Scapolite
Limonite	Scheelite
Lizardite	Schorl (Tourmaline)
Magnesite	Scolecite (Zeolite Group)
Magnetite	Serpentine
Melilite	Siderite
Mesolite (Zeolite Group)	Sillimanite
Microcline	Smectite
Monazite	Sodalite
Monticellite	Spessartite (Spessartine)
Montmorillonite	Spinel
Mullite	Spodumene
Muscovite	Staurolite
Natrolite (Zeolite)	Stilbite (Zeolite)
Nepheline	Stilpnomelane
Nephrite	Talc
Nosean (Sodalite Group)	Thomsonite (Zeolite Group)
Oligoclase	Titanite (Sphene)
Olivine	Topaz
Opal	Tremolite
Orthoclase	Tridymite
Palygorskite	Uraninite
Pectolite	Uvarovite

Alphabetical Index of Rock Forming Minerals cont...

Vermiculite	
Vesuvianite (Idocrase)	
Volcanic Glass (Obsidian)	
Wollastonite	
Zircon	
Zoisite	

Index of Rock Forming Minerals by Group

Alunite	Borates
Jarosite	Kernite
Alunite	Colemantite
Amphibole	Boracite
Ferrohornblende (Barkevikite)	Borax
Nephrite	**Brittle Mica**
Tremolite	Stilpnomelane
Anthophyllite	Talc
Grunerite	Pyrophyllite
Glaucophane	**Calcite**
Actinolite	Magnesite
Arfvedsonite	Siderite
Riebeckite (Crocidolite)	Calcite
Crossite	Carnotite
Cummingtonite	**Chlorite**
Hornblende	Clinochlore
	Chlorite
Andalusite	Chamosite
Andalusite (Chiastolite)	**Chromite**
Apatite	Chromite
Pyromorphite	**Clay**
Mimetite	Illite
Fluorapatite (Apatite)	Halloysite
Apatite	Kaolinite
Lazulite	Dickite
Aragonite	Palygorskite
Aragonite	Montmorillonite
Axinite	Smectite
Axinite	Vermiculite
Baryte	**Corundum**
Baryte (Barite)	Corundum
Celestite	**Dolomite**
Bauxite	Dolomite
Brucite	**Epidote**
Diaspore	Piedmontite
Boehmite (Bauxite Group)	Epidote
Gibbsite (Bauxite)	Pumpellyite-[Mg]
Beryl	Pumpellyite-[Fe]
Beryl	Clinozoisite

Index of Rock Forming Minerals by Group cont...

Allanite-[Y]	**Humite**
Allanite-[Ce]	Humite
Lawsonite	Chondrodite
Zoisite	
Evaporite	**Illite**
Halite	Illite
Carnallite	**Kaolinite**
Gypsum	Halloysite
Anhydrite	Kaolinite
Feldspar	Dickite
Plagioclase Group	**Kyanite**
Albite	Kyanite
Oligoclase	**Limonite-Goethite**
Andesine	Limonite
Orthoclase	Goethite
Bytownite	**Melilite**
Anorthite	Melilite
Sanidine	**Mica**
Microcline	Phlogopite
Labradorite	Fuschite
Feldspathoid	Pyrophyllite
Nepheline	Lepidolite
Sodalite	Muscovite
Leucite	Chloritoid
Cancrinite	Biotite
Fluorite	**Mineraloid**
Fluorite	Glass (Volcanic)
Garnet	Palagonite
Spessartite (Spessartine)	Leucoxene
Zircon	**Monazite**
Vesuvianite (Idocrase)	Monazite
Almandine	**Olivine**
Grossular (Grossularite)	Monticellite
Pyrope	Olivine
Uvarovite	Fayalite (Olivine)
Andradite	Forsterite (Olivine)
Titanite (Sphene)	**Opal (Hydrous Silicates)**
Staurolite	Opal
Graphite	**Oxides**
Graphite	Chromite
Hematite	Ilmenite
Hematite	Periclase

Index of Rock Forming Minerals by Group cont...

Pyroxene	Smectite
Enstatite	**Sodalite**
Spodumene	Nosean (Sodalite Group)
Diopside	Hauyne
Wollastonite	**Spinel**
Hedenbergite	Perovskite
Augite	Spinel
Jadeite	Magnetite
Rhodonite	**Topaz**
Bronzite	Topaz
Pectolite	**Tourmaline**
Acmite (Aegerine)	Schorl (Tourmaline)
Aegerine Augite	Cordierite
Hypersthene	Rubellite (Tourmaline)
Quartz	Dravite (Tourmailine Group)
Chalcedony	Elbaite (Tourmaline Group)
Quartz	**Uraninite**
Tridymite	Uraninite
Opal	**Vermiculite**
Cristabolite	Vermiculite
Rutile	**Zeolite**
Rutile	Heulandite
Cassiterite	Mesolite (Zeolite Group)
Scapolite	Erionite (Zeolite Group)
Scapolite	Scolecite (Zeolite Group)
Scheelite	Laumontite (Zeolite)
Scheelite	Analcite (Analcime)
Serpentine	Natrolite (Zeolite)
Chrysotile	Chabazite
Serpentine	Stilbite (Zeolite)
Antigorite	Thomsonite (Zeolite Group)
Prehnite	
Lizardite	
Sillimanite	
Sillimanite	
Mullite	
Dumortierite	
Smectite	
Montmorillonite	

References

Bayliss, P., Berry, L.G., Mrose, M.E., Sabina, A.P., Smith, D.K. (eds), 1983, "Mineral Powder Diffraction File". JCPDS, ICDD, 1005p.

Craig, J.R., Vaughan, D,J., 1981, " Ore Microscopy and Ore Petrography", John Wiley and Sons, New York, 406p.

Dana, J.D., "Manual of Mineralogy".

Deer, W.A., Howie, R.A. and Zussman, J, 1980, "An Introduction to the Rock Forming Minerals". Longman, London. 528p.

JCPDS-International Center for Diffraction Data, 1995, " Powder Diffraction File. Alphabetical Index, Inorganic Phases". JCPDS, International Center for Diffraction Data.

Jenkins, R., 1996." Introduction to X-ray Powder Diffractometry ". Wiley, New York.

Kerr, P.F., 1977, "Optical Mineralogy", McGraw-Hill, 492p.

Le Maitre, R.W.(ed), 1989, "A Classification of Igneous Rocks and Glossary of Terms". Blackwell Scientific Publications, Oxford, UK. - Igneous plots and systematics.

Myashiro, A.,1973. "Metamorphism and Metamorphic Belts". Allen and Unwin, London,492p.

Mutschler, F.E., Rougon, D.J., Lavin, O.P., Hughes, R.D., 1981, "Petros - A Data Bank of Major Element Chemical Analyses of Igneous Rocks for Research and Teaching (Version 6.1)". NOAA-National Geophysical and Solar-Terrestrial Data Center.

Nichols, M.C., Nickel, E.H., 1991, "Mineral Reference Manual", Chapman and Hall, New York, 250p.

Wills, B.A., 1992, "Mineral Processing Technology". Pergamon, Oxford, 855p.

Web Sites and Software:

Athena Mineralogy http://un2sg1.unige.ch/www/athena/mineral/mineral.html - List of Mineral Names, Mineralogy

Geologynet.com http://www.geologynet.com/indexa.htm - Mineralogy

LR Ream Publishing http://www.mineralnews.com/ - Publishers of The Mineral Database (TMD)

Mineral Database - Materials Data Mineral Database, Formerly Aleph Interprises

Mineralogy Database http://webmineral.com/ - HTML Mineral Database

MinDat.org http://www.mindat.org/ - Mineralogy Database

Picture Credits:

Cover photo by the author

www.ingramcontent.com/pod-product-compliance
Lightning Source LLC
Chambersburg PA
CBHW081119170526
45165CB00008B/2491